International Approaches to Professional Development of Mathematics Teachers

International Approaches to Professional Development of Mathematics Teachers

Edited by
Nadine Bednarz, Dario Fiorentini, and
Rongjin Huang

University of Ottawa Press
Ottawa

University of Ottawa Press
542 King Edward Avenue
Ottawa, ON K1N 6N5
www.press.uottawa.ca

uOttawa

The University of Ottawa Press acknowledges with gratitude the support extended to its publishing list by Heritage Canada through its Book Publishing Industry Development Program, by the Canada Council for the Arts, by the Canadian Federation for the Humanities and Social Sciences through its Aid to Scholarly Publications Program, by the Social Sciences and Humanities Research Council of Canada, and by the University of Ottawa.

LIBRARY AND ARCHIVES CANADA CATALOGUING IN PUBLICATION

International approaches to professional development for mathematics teachers / edited by Nadine Bednarz, Dario Fiorentini and Rongjin Huang.

Includes bibliographical references and index.
ISBN 978-0-7766-0747-4

1. Mathematics--Study and teaching. 2. Mathematics teachers--Training of. I. Bednarz, Nadine II. Fiorentini, Dario III. Huang, Rongjin

QA11.2.I58 2011 510.71 C2011-902965-0

Contents

Part 3
Professional Knowledge Development of Mathematics Teachers

Part 4
Experiences of Professional Development in Teaching Mathematics

Introduction

Experiences in Different Countries in the Professional Development of Mathematics Teachers

Nadine Bednarz, Dario Fiorentini, and Rongjin Huang

This book originates from the discussions of an international study group during the eleventh International Conference on Mathematics Education (ICME-11) in Monterrey, Mexico, from 6 to 13 July 2008. It is the result of a long process coordinated by the three editors of organizing the different papers presented at this conference to clarify the experiences in different countries in the education of practising teachers of mathematics.

The chapters offer different international perspectives on this topic. They come from researchers involved in the field of mathematics teacher education in Brazil, Canada, China, Mexico, Israel, Portugal, the United Kingdom, and the United States and present, in various contexts, different tendencies, theoretical perspectives, and experiences, raising questions and issues confronting mathematics teacher educators and researchers. The chapters together put forward some answers to the following main question: What do we know about the experiences and approaches developed in different countries to support the professional development of teachers *for* practice, *in* practice, and *from* practice?

Overview of the Book

The interest in mathematics teacher education in the international community is relatively recent (see Jaworski, this volume). As Jaworski shows, the research in mathematics education during the past twenty years has come progressively to a recognition of mathematics teacher education as a fundamental subfield (offering an increasing number of publications, researches, and communications). In particular, the education of practising mathematics teachers appears to be an important preoccupation of the international community. The trajectory of this subfield of research shows major shifts in its evolution: from research *on* teachers (e.g., centred on their beliefs, knowledge, and conceptions) to research *from* teachers, informed by them and their practice (e.g., learning from teachers about their professional knowledge, practice, and gestures) to research *with* teachers (e.g., contributing to the construction of new knowledge related to mathematical teaching). This book mirrors these contemporary tendencies.

Indeed, the professionalization of teaching and teacher education, along with the complexity of mathematical teaching practices, raises questions for teacher educators and researchers in relation to the education of practising teachers. Nowadays, schools are confronted with many challenges, such as the elaboration of new curricula grounded in learning/teaching paradigms that call into question a number of mathematical teaching practices, the consideration of specific cultural contexts requiring a rethinking of curricula and mathematical practices, the integration of cultural/historical dimensions into teaching, the development of interdisciplinary projects, the introduction of new technologies into classrooms, and the adaptation of teaching practices to take account of different students and contexts (e.g., students with learning difficulties, multicultural classrooms, ESL/teaching mathematics in a second language, underprivileged schools, adults, and analphabetism). The problems encountered differ according to the teacher education systems in place in each country.

These challenges demand serious reflection on how to support those who are directly concerned, the practising teachers and other school practitioners, and how to develop means that take into account the differing issues related to the continuous education of teachers in different countries and settings. The education of practising teachers can take different forms (e.g., courses, programs, shared experiences, professional development through professional life, teachers as researchers, action research, and collaborative research) and can be supported by different persons (e.g., teachers, pedagogical consultants, teacher educators, researchers, and mathematicians).

In the past, in-service education initiatives were often considered inappropriate or too distant from real classroom practices and teachers' knowledge to be effective for and supportive of teachers. However, in the past decade, different experiences and alternative approaches have been developed worldwide to support the learning process of practitioners from, for, and in practice (e.g., web communities, mentoring, action research, collaborative research, communities of inquiry, and teachers as researchers). Most of these approaches and experiences take seriously practitioners' points of view and knowledge in supporting and enhancing teachers' practice, mainly through the exploration of mathematical content taught (exploring content in relation to practice and enriching it), the construction of mathematical professional knowledge linked to teaching practices, the construction of learning/teaching situations in mathematics (working with teachers on the elaboration and analysis of these situations in terms of specific contents or processes), or the undertaking of investigative activities in mathematics.

This book offers a portrait of these different approaches and experiences, rooted in practice, informed by this practice, and working in return for this practice. These approaches/experiences aim to support the professional development of teachers and the production of a significant body of knowledge in mathematics education related to practice. The chapters offer research, new theoretical perspectives, and reflections and practical experiences from around the world to support the professional development of mathematics teachers.

The book addresses a number of questions.

- What do we mean by the professional development of mathematics teachers?
- What conceptions of professional development are emerging from the research and experience in mathematics education? What are their different theoretical and epistemological bases?
- How does the professional life of a teacher contribute to his or her professional development?
- What do we know about the different approaches and experiences developed to support the professional development of teachers? How are they characterized?
- Especially, how can we characterize the process developed within and throughout these approaches and experiences in regard to their relation to practice, to the role of teacher educators and teachers, and to the learning taking place?
- What strategies, practices, and processes are identified and described as promoters or catalysts of teachers' professional development?
- What do we know about the sorts of professional learning taking place in these approaches and settings?

- How does the education of practising teachers address the complexity of practice in mathematics education and the challenges with which it is confronted (e.g., implementation of reforms or changes in curriculum)?
- How can the process of professional development be thought of differently at the macrolevel?

The chapters offer various answers to these questions, shedding light on innovative approaches to the professional development of mathematics teachers.

Content of the Book

The chapters are organized into five parts, which constitute complementary aspects of the problematic of the education of practising mathematics teachers.

Over the past two decades, as discussed in Part 1, mathematics teacher education has become an important subfield of mathematics education. Jaworski traces the historical genesis of this development, highlighting key elements. Her chapter addresses the current major trends in the education of practising mathematics teachers. The theoretical bases, the different approaches developed in relation to the teacher's role, the learning taking place, and the development of mathematics teacher educators are her focus. Her chapter ends with a perspective on the future of mathematics teacher education, presenting some key issues and questions.

Part 2 addresses the professional development of mathematics teachers in practice. We often think of formal approaches (programs, courses, etc.) or informal experiences organized by mathematics teacher educators or researchers to support this professional development. However, the professional life of a teacher is often an occasion, in itself, of professional development. The professionalization of teaching requires teachers to be involved in a learning process throughout their professional lives. Several researchers have focused on the cycles of the professional lives of teachers (Huberman, 1989; Katz, 1986) or on the entry into the profession of beginning teachers (Kagan, 1992; Louvet & Baillauquès, 1992). In fact, rich experiences of mathematics teachers during their careers can contribute to their professional development. It is important to analyze these experiences to better understand their possible contributions and the learning processes taking place in these cases. The three chapters comprising this second part are situated in that perspective. Different professional trajectories are described and analyzed, opening up new and significant avenues to support teachers in the process of life-long learning and professional development. They concern the change of role during the career, from that of teacher (working with students) to that of subject leader/pedagogical consultant (working with other teachers) (see Nunes); the mentoring of pre-service and beginning teachers by experienced practising teachers (see Rhodes & Wilson); and the description of a professional system in support of professional development (see Li et al.). All these professional events constitute rich opportunities. What learning is taking place in each case? What characterizes these different professional opportunities?

As discussed in Part 3, research in the international community has pointed to the specificity of mathematics for teaching. How mathematics teachers engage with a mathematical situation in their teaching is quite different from how mathematicians did it (Ball & Bass, 2003). One issue that has received recent attention is the gap between the mathematical experiences encountered in university courses/in-service education and the practice of teaching mathematics in schools (Proulx & Bednarz, 2008a, 2008b). This gap has been characterized in terms of the mathematical content explored in each case, academic courses focusing on formalism, compacted ideas, and abstract forms, mathematics teaching experiences in which decompacted ideas are necessary to support a deep understanding of concepts by students (see Ball & Bass, 2003; Moreira & David, 2008). This gap has also been related to how this content is approached (see Bauersfeld, 1994; Burton, 2004). This divide raises important questions about the current orientations of mathematical preparation in teacher education programs and suggests a rethinking of the mathematical experiences and learning opportunities that teachers are exposed to in their professional formation.

Some researchers in the international community have questioned and rethought the mathematical education provided to mathematics teachers (at the pre- and in-service levels) (see Proulx & Bednarz, 2008a, 2008b, for more details). The three chapters in Part 3 of the book offer teachers rich mathematical experiences developed from that perspective.

In the chapter by Proulx, research interests are located in studying an approach to the education of practising teachers that attempts to offer them experiences better aligned with their mathematics teaching practices: that is, an initiative focused on the development and enrichment of teachers' knowledge of the mathematics that they teach—in other words, school mathematics. In the chapter by Sanchez, the initiative described offers an engaging practice to teachers in which they are mathematical "doers." This practice, called a note of reflection, is a fictitious classroom event arranged around a mathematical question or activity. The analysis is centred on the dialogue between teachers emerging from such an experience. The chapter by Murphy focuses on the learning of advanced mathematical content by teachers who have to teach at an advanced level. Several researches have pointed out the rupture between the culture of secondary mathematics and the culture of postsecondary advanced mathematics, leading to learning difficulties by students in the passage from one level to the other (Corriveau & Tanguay, 2007; Luk, 2005). This rupture is characterized, for example, by increasing requirements in terms of the use of mathematical language and symbolism, a higher level of abstraction, and higher demands on the use of proofs. The initiative developed by Murphy is aimed at teachers in secondary schools to prepare them to teach advanced mathematical content at a higher level. The course focuses on this advanced mathematical content in alternation with practical teaching experiences. Assessments developed in the course consider various rich dimensions: essays focusing on one aspect of mathematical content knowledge, classroom-based inquiries involving research, and mathematical investigations in which participants reflect on the process of working on mathematics and the implications that this has for their teaching.

What we can call professional mathematics (Proulx, Bednarz, 2009, 2011), mathematics more aligned with teachers' professional practices, appears, in these different chapters, as a point of entry in engaging teachers in professional development.

In Part 4, alternative approaches have been developed by researchers or teacher educators to support the learning process of practitioners and overcome the gap between research and practice. Different avenues are illustrated and analyzed here. Some of them take the form of formal experiences developed at the university (programs or courses), while others describe informal experiences with teachers in practice. Different underlying models and theoretical bases are developed.

In some cases, models are developed a priori by researchers alone, on the basis of previous research in a certain field. These models are then used to design a professional development approach for teachers and improved to analyze the learning taking place. This is the case of two chapters in this fourth part. The Beatty and Bruce chapter centres on a research/professional development model, based on previous work on algebra and patterning, and an assessment of this model, showing its influence on the development of pedagogical content knowledge by the teachers involved in it. The Kramarski and Revach chapter focuses on a self-regulated research model and analyzes its influence on how teachers construct knowledge. In these two cases, teachers are considered as students and participants (see Jaworski, this volume).

In the six other chapters in Part 4, authors describe models of research/ professional development from a participatory perspective, working *with* teachers more than *on* teachers (Bednarz, 2004; Desgagné & Bednarz, 2005; Lieberman, 1986). From this perspective, approaches to teacher education and research are articulated on practitioners' teaching practices and take account of their voices, their points of view, their levels of professional knowledge, their routines of action, the meanings that they give to teaching or learning situations in mathematics. In the chapter by da Costa, the underlying model guiding the teacher educator/researcher is Schön's (1983) model of reflection in action and on action, teacher educators favouring the reconstruction (by teachers) of situations in which actions are produced, with their underlying assumptions (what the author calls practical reflection for social reconstruction). We addressed here statistical content knowledge (using computers) in which teachers are not comfortable. The change in roles during the process (from teachers as students, to teachers in classrooms, to teacher educators or researchers) contributes to the professional development of teachers.

This mediation between different roles and different cultures is also at the centre of the chapter by Nacarato and Grando, where graduate students (preservice teachers), postgraduate students, practising teachers, and university educators/researchers work together on the elaboration of teaching situations in geometry using different tools/media. This formative process (for all the participants) is the result of "a sharing/learning experience in geometry in a collaborative group composed of different actors," leading to a rethinking of geometrical school concepts and mathematical classroom cultures.

In these two chapters (da Costa, Nacarato, and Grando) as in the next one (by Ferreira & Miorim), the idea of a collaborative group is central. As we can see in the Ferreira and Miorim chapter, which analyzes the evolution of such a collaborative group (composed of a researcher and teachers) during one year, this shared experience contributes both to the professional development of teachers and to the learning of teacher educators/researchers.

The other three chapters are more related to what we can call investigative/inquiry activity and reflection. In the first case (the chapter by Suurtamm & Vézina), the model of professional development is designed to provide teachers with experiences to support them as they move from a model of teachers' explanation to that of listening to students' mathematical thinking in problem solving. In the second case (the chapter by Passos & Lamonato), the underlying model guiding the researchers is based on the fact that a mathematical investigation can trigger discussion, problematization, and enlargment of teachers' knowledge. In the course developed in geometry by the two educators for preschool teachers, teachers have to experiment with mathematical investigative activities, reinvest them in their practice, and reflect on them. Investigative activities are also at the core of the last chapter in Part 4, by Freitas and Fiorentini. Their chapter explores the potentialities of investigative activities and of the writing process in two different courses, a geometry course for pre-service teachers and a teaching mathematical course involving practising teachers in the supervision of future teachers. Writing here appears as a particular way to reflect on their practice, to reconstruct this practice and its underlying rationality, contributing to the professional development of teachers and the construction of a professional identity by future teachers.

Part 5 turns to the fact that schools today are confronted with important reforms in education. The implementation of these reforms and curricular changes is often conceived of as a "top-down" process in which the orientations and programs are built by a group of experts and then presented to teachers so that they can apply them. Sessions are organized around objectives, textbooks, and materials, and teachers are considered as students in order to reappropriate new curricula, materials, textbooks, and assessments in their practices. This top-down model of development has been questioned and criticized by several researches, who have shown that this process has no influence on mathematical teaching practices in schools. These critics have rethought the process of development of curriculum in schools. How can we support the people directly concerned with these curricular changes, taking account of the different issues that they encounter, of their professional knowledge, of their practice? The first two chapters offer new ways of approaching this important question. In the first chapter (by Chissick), the researcher explores factors affecting the implementation of reforms in secondary school mathematics, from the inside, from the point of view of teachers. In the second chapter, Fiorentini et al. describe a research program centred on different experiences of teachers taking responsibility for curriculum development better aligned with their practice. The relationship between this curriculum development by teachers and their professional development is documented.

At the end of Part 5, the definition of continuous professional development at a macrolevel is considered. In their chapter, de Geest et al. offer a new way to approach the conceptualization of professional development (usually the responsibility of researchers alone). They propose a new way to design continuous professional development to overcome the gap between research and practice, involving at different stages different actors concerned with this question.

The rich portrait drawn by these different chapters on the education of mathematics teachers informs the community of teachers, teacher educators, and researchers on new alternatives in mathematics teacher education and on important developments in this field.

References

Ball, D. L., & Bass, H. (2003). Toward a practice-based theory of mathematical knowledge for teaching. In E. Simmt & B. Davis (Eds.), *Proceedings of the 2002 annual meeting of the Canadian Mathematics Education Study Group* (pp. 3-14). Edmonton: CMESG.

Bauersfeld, H. (1994). Réflexions sur la formation des maîtres et sur l'enseignement des mathématiques au primaire. *Revue des sciences de l'éducation, 20* (1), 175-198.

Bednarz, N., Proulx, J. (2011). An attempt at defining teachers'mahematics of their practice through research on mathematics at work. Proceedings of the 7th European Conference of Research in Mathematics Education (CERME), Rzeszow, Poland, 9-13 février, 10 pages.

Bednarz, N., Proulx, J. (2009). Knowing and using mathematics in teaching: Conceptual and epistemological clarifications. *For the Learning of Mathematics*, 29 (3), november, 11-17.

Bednarz, N. (2004). Collaborative research and professional development of teachers in mathematics. In M. Niss & E. Emberg (Eds.), *Proceedings of the International Conference on Mathematics Education* (ICME-10), 4-11 July. Copenhagen: Cédérom.

Burton, L. (2004). *Mathematicians as enquirers: Learning about learning mathematics*. Dordrecht: Kluwer.

Corriveau, C., & Tanguay, D. (2007). Formalisme accru du secondaire au collégial : Les cours d'algèbre linéaire comme indicateurs. *Bulletin de l'Association Mathématique du Québec, 48*(1), 6-25.

Desgagné, S., & Bednarz, N. (2005). Médiation entre recherche et pratique en education : Faire de la recherche "avec" plutôt que "sur" les praticiens. *Revue des sciences de l'éducation, 31*(2), 245-258.

Huberman, M. (1989). *La vie des enseignants : Evolution et bilan d'une profession.* Neufchâtel: Delachaux et Niestlé.

Kagan, D. (1992). Professional growth among preservice and beginning teachers. *Review of Educational Research, 62*(2), 129-169.

Katz, L. G. (1986). Developmental stages of preschool teachers. In J. Nias (Ed.), *Teacher socialization: The individual in the system* (pp. 56-59). Deakin, Australia: Deakin University Press.

Lieberman, A. (1986). Collaborative research: Working with, not working on ... *Educational Leadership, 43*(5), 29-32.

Louvet, A., & Baillauquès, S. (1992). *La prise en fonction des instituteurs.* Paris: Institut National de Recherche Pédagogique.

Luk, H. S. (2005). The gap between secondary school and university mathematics. *International Journal of Mathematical Education in Science and Technology, 36*(2-3), 161-174.

Moreira, P. C., & David, M. M. (2008). Academic mathematics and mathematical knowledge needed in school teaching practice: Some conflicting elements. *Journal for Mathematics Teacher Education, 11*(1), 23-40.

Proulx, J., & Bednarz, N. (2008a). Re-conceptualizing the mathematical preparation of secondary mathematics teachers. *Alberta Journal of Educational Research, 54*(1), 124-127.

Proulx, J., & Bednarz, N. (2008b). The mathematical preparation of secondary mathematics schoolteachers: Critiques, difficulties, and future directions. Paper presented at ICME-11 in TSG-29, Monterrey, Mexico. Retrieved 16 March 2009 from http://tsg.icme11.org/tsg/show/29.

Schön, D. A. (1983). *The reflective practitioner: How professionals think in action.* New York: Basic Books.

Part 1

International Perspectives on Mathematics Teacher Education

Chapter 1

Situating Mathematics Teacher Education in a Global Context

Barbara Jaworski

Introduction

This chapter stems from an invited presentation at ICME-11 in Topic Study Group 28 focusing on the in-service education, professional life, and development of mathematics teachers. I was to set the scene historically and offer my own perspective, from an international point of view, on where we currently are in this field. I welcomed the invitation as it gave me the chance to reflect on my own years of work in this field and share my reflections with others. What I present here is a personal account.[1] I have tried to be as comprehensive as possible where major events are concerned but cannot guarantee that I have not missed something. I am also strongly aware that my account comes from an English-speaking environment and that language denies me access to events and publications in which the language is not English. However, the rest of this book includes chapters from diverse settings that reflect research activities in a wide range of nations.

Structure of the Chapter

This chapter is composed of three parts: (1) a historical perspective on the (sub)field of mathematics teacher education; (2) some themes in mathematics teaching development; and (3) issues in mathematics teaching development and one way forward.

I consider mathematics teacher education to be a subfield of the discipline of mathematics education. I think that history supports this claim.

I therefore set out what I see as key aspects of the development of this subfield in terms of international events and publications. From there I move into the "content" of the subfield. Here I limit myself mainly to the topic of the study group, which concerns the education and development of *practising* teachers. In the main, this chapter is not about initial or pre-service teacher education, the education of new teachers; that in itself has a considerable literature. I offer what I consider to be main themes in the development of the education of practising teachers. Finally, I recognize a range of issues relating to teacher learning and teaching development in mathematics and offer, tentatively, what I see as being one way forward.

To avoid using the words *mathematics teacher education* repeatedly, I will sometimes use the acronym MTE. Also, MTEs are mathematics teacher educators.

A Historical Perspective on Mathematics Teacher Education

Here I trace the development of our subfield through major events and contributions internationally over the past twenty years. One of the major landmarks for mathematics educators over the years is the annual PME conference (Conference of the International Group for the Psychology of Mathematics Education[2]), dating back to 1976. I start with the development of MTE as seen through the PME conferences and their proceedings during this time and continue to consider other key events and related publications.

A Reconstruction of the Evolution through PME

I trace events from 1986, the year that I attended my first PME conference. Celia Hoyles (1992) refers to the earlier PME conferences, suggesting that the majority of papers in these early days focused on student understanding of mathematical concepts. Regarding 1979 (PME 3), she writes that "if the teacher was mentioned at all, s/he was purely a facilitator—to dispense facts and information; to identify errors or misunderstandings; to provide materials or strategies to overcome misconceptions; or to promote further mathematical development" (p. 3: 263). It seems to me that the nature of being "purely a facilitator" lies more in how such aspects of role are treated than in the aspects themselves. A suggestion is that perceptions of the role of the teacher at this time were underdeveloped, that the teacher was almost dismissed in the simplicity of the conception of teaching, and that overcoming misconceptions or promoting development was seen in instrumental ways.

Between 1979 and 1985, Hoyles found a few papers that started to refer to the teacher's role in the classroom and to consider teachers' beliefs and their relationship with teaching practice. However, in 1986, there was just one paper explicitly related to teaching: a case study of the socialization of one teacher in his first year of teaching. During the next two decades, a research focus on teachers' beliefs continued alongside other factors relating to teachers and teaching.

Table 1.1 provides a brief overview of development from 1986.

Table 1.1
Twenty years of PME conferences 1986–2006

PME #	Year	Papers and their content and other key events
PME 10	1986	1 paper relating to teaching
PME 11	1987	19 papers in themes headed as instruction and teacher training
PME 14	1990	No thematic categories designated as teaching or teacher education Papers on teaching or learning to teach spread throughout other categories Three working groups in areas of mathematics teacher education
PME 15–27	12 years	Development of mathematics teaching and teacher education as major research themes Two plenary presentations surveying the subfield Three books arising from working groups at PME
PME 28	2004	50 papers in themes headed teacher classroom practice, teacher education and professional development, and teacher knowledge
PME 30	2006	27 papers in themes on in-service teacher development, pre-service teacher development (elementary), teacher content knowledge, teacher pedagogical knowledge, and teacher thinking A review of 30 years of PME

PME conference proceedings contain a thematically organized list of the papers presented at a conference. The titles given to these themes are themselves illustrative of thinking at that time. For example, in 1987, we see the thematic titles "Instruction" and "Teacher Training." More recently, the words *instruction* and *training* have been subject to critique in terms of their perhaps behaviourist connotations, that pupils are to be *instructed* and teachers *trained* according to some given conventions that are the province of experts in the field. The shift in thematic titles to "Teacher Classroom Practice," "Teacher Education and Professional Development," and "Teacher Knowledge" in 2004 also reflects a shift in the discourse of teacher education. As more consideration was given to the role of a teacher and the nature of teaching, earlier, more simplistic, notions gave way to wider recognition of context and complexity in teaching. The discourse changed accordingly. In 2006, moreover, the thematic categories emphasized differences between "Pre-Service and In-Service Development" and "Teacher Knowledge and Teacher Thinking." We see here a maturing within the subfield: finer categorization of papers, more refined discourse, deeper consideration of factors and issues in teaching and its development with relation to classrooms at different levels, and associated education of teachers. Papers classified in these sections number about half of those classified in the subfield in 2004, partly due to new thematic categories in PME, such as "Classroom Culture" and "Social Activity Theory," which dealt with teaching but whose focus was methodological, institutional, or theoretical. Overall, however, there were fewer papers in 2006, so it remains to be seen whether this is a downward trend.

Alongside developments in the research reported at PME, there was such interest in mathematics teacher education by 1990 that three working groups had been formed. They were known respectively as

1. Psychology of In-Service Education of Mathematics Teachers (a research perspective);
2. Research on Psychology of Mathematics Teacher Development; and
3. Teachers as Researchers in Mathematics Education.

The first focused on in-service teacher education, what we now call education of practising teachers, the substance of this book. The second looked at teacher education and development more broadly, and the third focused specifically on teachers' engagement in research. These three working groups continued well into the 1990s with varying membership and leadership. As the work of each group came to an end, it produced a book of papers relating the work of the group over the years. The first volume to be published (Zack, Mousley, & Breen, 1997: Group C) focused largely on teachers' engagement in inquiry and research and corresponding developments in teaching practice; the second volume (Jaworski, Wood, & Dawson, 1999: Group A) concentrated on in-service teacher development; and the third volume (Ellerton, 1999: Group B) focused on both pre-service and in-service development. Looking back on these books, I believe that they offer an important record of the thinking and practice in mathematics teacher education at that time. It was an interesting and in many ways exciting time. Constructivism largely held sway as a basic theoretical perspective (Davis, Maher, & Noddings, 1990), and many of the chapters discussed teacher education initiatives that were rooted in constructivist principles and associated methodology (see also Confrey & Kazak, 2006). We were moving away from visions of mathematics teaching as direct instruction and toward ideas related to problem solving and more creative approaches involving students in mathematics (Grouws & Cooney, 1988; Romberg & Carpenter, 1986). As we thought about student involvement in exploring mathematics, we saw the beginnings of involving teachers in exploring mathematics teaching (ATM, 1987; Krainer, 1998). The roles of teacher educators-researchers in this process started to come into focus (Jaworski, 1994, 1998). The book's chapters offer a panorama of MTE activity related to this landscape.

In the period from PME 15 to 27, two conferences are of particular note (PME 16, 1992; PME 18, 1994), since each had a keynote presentation focusing on mathematics teacher education and offering a survey of research to that time, the first from Celia Hoyles and the second from Joao Pedro da Ponte (Hoyles, 1992; Ponte, 1994). These two papers together chart a progression in ways of seeing teachers and teaching from a one-dimensional, largely instrumental, focus to research in which interpretations seek to account for multiple dimensions and serious complexity in teaching.

In 1992, based on papers at PME up to that time, Hoyles (1992) suggested that up to 1987 the teacher was considered hardly at all other than as an adjunct of students' learning of mathematics, "a passive conveyer of facts and information" (p. 3: 625). According to Hoyles, early papers on teachers and teaching reflected the teacher's focus on curriculum delivery and differentiation of students according to ability. Many papers focused on the beliefs of teachers: "there appears to be an implicit view that teachers have 'beliefs' in some decontextualised sense which need to be accessed and changed" (p. 3: 266). The focus on beliefs led to suggestions of "true beliefs which may not be enacted in practice" (p. 3: 265). Hoyles had perceived a strong focus on the influence of teacher beliefs on classroom activity and perceptions of a gap between espoused and enacted beliefs. She reported on research that saw inconsistencies between teachers' beliefs and their practice as obstacles to the success of in-service development and therefore as something to be changed. This suggested a need to investigate the interaction of teacher beliefs and curriculum innovations. Hoyles marked a shift in the late 1980s toward a focus on social interaction and construction in which teachers' classroom activities were heavily influenced by classroom and school cultures and those of the wider society. Methodologically, the period marked a shift to more qualitative research approaches, particularly case studies, that "illuminate the general through the particular" (p. 3: 282) and go some way toward addressing context and complexity while presenting problems in critical judgment of interpretations.

Two years later, Ponte (1994) examined the place of the teacher in mathematics education research, particularly mentioning negative perspectives—seeing the teacher as an instrument, as a deficient professional, as "a person with deep misconceptions, lack of mathematical knowledge, and inappropriate and inconsistent beliefs, contradictory with current reform efforts" (p. 198). Ponte recognized research that focuses on teachers' knowledge of mathematical concepts and how to teach them, the assumption being that teachers who do not know their subject cannot do a good job in teaching it. He also pointed to more didactically oriented studies focusing on subject matter and pedagogical knowledge but lacking a clear view of how such knowledge develops and works in practice. Hoyles (1992) had pointed toward teacher research as an emergent focus; Ponte acknowledged research on reflective practice and teachers as researchers, seen as playing an important role in defining the purposes and goals of their work as well as attaining them. Ponte elaborated on the issues raised in his review of research through three short accounts of teachers' conceptions in relation to mathematical problem solving in the Portuguese curriculum. Although the accounts raised issues relating to teachers' knowledge and beliefs in much of the research reviewed, Ponte's analysis took into account social and institutional factors, emotional responses, and self-image. Ponte considered different worlds of experience that give rise to different interpretations, according to changes in context, and different forms of knowing, including knowing in action in contrast to definitive or declarative knowledge. The picture painted in these accounts is one of deep complexity in which research observations are subject to interpretation relative to theoretical and socio-cultural factors, which need to be made explicit to make sense of what is observed. Overall, there emerges a deep understanding of the complexities that teachers face and the importance of research giving teachers a voice—a stark contrast to the deficiency perspective presented earlier.

In 2006, in a celebration of thirty years of PME, a specially commissioned volume documented research in a number of key areas, one of which was mathematics teacher education. Two review chapters were included: the first from Salvador Llinares and Konrad Krainer and the second from Joao Pedro da Ponte and Olive Chapman (Llinares & Krainer, 2006; Ponte & Chapman, 2006). They bring us approximately to the present in terms of perspectives and issues in the subfield. I will refer to these papers below.

A New Journal in the Field of Mathematics Teacher Education

In parallel with preparation of the books mentioned above, in 1998 a new journal was launched, the *Journal of Mathematics Teacher Education.* It came about from an agreement between the publisher, Kluwer, and Professor Tom Cooney at the University of Georgia, United States, to advertise the publication of a book or books in mathematics teacher education. A call for papers was made, and by the closing date more than 100 papers had been submitted. This quantity suggested the viability of a journal, and a decision was made to launch *JMTE,* with Cooney as the founding editor. To date, the journal has flourished, with increasing numbers of issues per year and new editorial teams.[3]

The mission statement of *JMTE* is as follows:

> The *Journal of Mathematics Teacher Education* (*JMTE*) is devoted to research that seeks to improve the education of mathematics teachers and develop teaching methods that enable mathematics students to learn better. The journal covers all stages of the professional development of mathematics teachers and teacher-educators. It serves as a forum for examining institutional, societal, and cultural influences that impact on teachers' learning and ultimately their students' learning.[4]

JMTE initially invited papers of two kinds:

1. research papers relating to programs for educating prospective teachers of mathematics, programs for educating practising teachers of mathematics, programs in mathematics teaching development, and theoretical perspectives in mathematics teacher education; and

2. papers relating to mathematics teacher education around the world.

Cooney had the foresight to realize that a journal in English, edited from the United States, might have difficulty attracting papers from a truly international constituency. The section on MTE around the world was designed to provide insight into programs in teacher education worldwide. Nevertheless, throughout ten years of the journal, *JMTE* papers came overwhelmingly from the United States, with a relatively small number coming from a range of other countries. This was despite an international editorial board and requests to members to encourage papers from their own countries. I take up this issue again below.

JMTE has not had a thematic organization of papers: thus, it would be a major task to trace themes through the papers published (though it could be an interesting task). However, a simple count shows that a little more than 10% of papers were published in the category of MTE around the world. During the first ten years of *JMTE,* three special issues were published, one on "Communities in Mathematics Teacher Education" (*JMTE 6*(2), 2003); one on "Inter-Relating Theory and Practice in Mathematics Teacher Education" (*JMTE 9*(2), 2006); and a triple issue on "The Nature and Role of Tasks in Mathematics Teachers' Education" (*JMTE 10*(4, 5, 6), 2008). In the main, during the first decade of the journal, the editors concentrated on establishing *JMTE* as a major academic journal with a high scientific quality of papers. Doing so had negative associations also, as we shall see below. Currently (in 2008), the journal is anticipating further special issues and moving in new directions. There is a need to find ways to create more opportunities within the international community: for example, national portraits on mathematics teacher education, presentations of research tendencies, results, theoretical perspectives, and issues that could regroup researchers in the subfield internationally.

Mathematics Teacher Education in International Settings

Evidence of activity in mathematics teacher education internationally can be found in a range of other events and publications.

The four-yearly conferences of the International Congress of Mathematics Education (ICME) have always included working groups focusing on different aspects of teacher education, including a group on in-service teacher education and the professional lives of mathematics teachers. Proceedings from these conferences include brief details from the work of these groups. This book, edited from Canada and Brazil, is the result of such a group meeting at the most recent conference, ICME-11, in Monterrey, Mexico.

ICME conferences are organized by the International Commission on Mathematical Instruction (ICMI), which celebrated 100 years in 2008. Celebrations included a special conference organized in Rome with one group focusing on mathematics teacher education (Grevholm & Ball, 2008).

In Europe in 1997, a European Society for Research in Mathematics Education (ERME) was established. Its congresses, CERME, have been held every two years since 1999. These conferences are based on working groups in key areas of mathematics education chosen by the respective program committee. Mathematics teacher education has been the topic of one group at each of the conferences so far and, in each case, has been one of the largest groups in terms of papers submitted and numbers participating. The group at the first conference produced a special volume of papers relating to research in MTE focusing on issues raised at the conference (Krainer, Goffree, & Berger, 1999). Papers from succeeding conferences can be found in the published proceedings of each conference.

In addition to the regular conferences of established groups, other one-off conferences in mathematics teacher education have been organized internationally, often with invited participants: for example, in Taipei, Taiwan, in 1999, from which a book, *Making Sense of Mathematics Teacher Education* emerged (Lin & Cooney, 2001); in Malmø, Sweden, in 2003, proceedings of which, *Educating for the Future,* were published (Straesser, Brandell, Grevholm, & Helenius, 2004); and in Oberwolfach, Germany, in 2007, with a published report including abstracts of presented papers (Reiss, Schoenfeld, & Törner, 2007). Thus, from these events we see a growing literature base in MTE.

A Special Survey

In preparation for ICME-10 in 2004, a survey was commissioned by ICMI on research in mathematics teacher education. As a result of their literature searches, the researchers identified a number of key issues and made four major claims (Adler, Ball, Krainer, Lin, & Novotna, 2005). The claims are summarized here as follows (adapted from Adler & Jaworski, 2009):

1. Small-scale qualitative research predominates. This most often involves research that focuses on a single teacher or on small groups of teachers (n<20) within individual programs or courses.

2. Most teacher education research is conducted by teacher educators studying the teachers with whom they are working. Most studies are small case studies in which researchers have some direct involvement and thus some interest in the case being studied.
3. Research in countries where English is the national language dominates the literature surveyed. In *JMTE* between 1998 and 2003, this was 80% of published papers, in *Journal for Research in Mathematics Education* (*JRME*) 71%, and in PME, between 1999 and 2003, 43%.
4. Questions that come to constitute the research field are driven by concerns in particular contexts and thus might not reflect the diversity of problems in teacher education that exist globally. Some questions have been studied extensively, though other important questions remain unexamined. Much of the research, particularly in the United States, has been concerned with reform and has involved efforts to show that particular programs of teacher education "work." As a consequence, for example, we know much less than we should about teachers' learning from experience: whether teachers learn, what they learn, and what supports learning from experience.

These findings were salutary for the subfield in general and for *JMTE* in particular. Issues are discussed in Adler et al. (2005) and Adler & Jaworski (2009). Briefly here, I see the issues as centring on internationalism in MTE. The language issues noted, the domination of the English-speaking world in published papers, and the domination of the reform movement in the United States on issues in the subfield raise serious concerns for the subfield.

The ICMI Study 15

Since 1990, ICMI has developed a series of studies in specially designated areas of research in mathematics education. In the first study, on mathematics and cognition, Balacheff (1990), in collaboration with a team of researchers from PME, wrote as follows: "What we know virtually nothing about is the interactive nature of teachers' and students' conceptions: how one influences the others, their origins, the intensity with which they are held, and how permeable they are in the face of classroom dynamics" (p. 141). They wrote further that "what we have developed here confirms the importance of studies taking the teaching situations as a research object:

how they function with the purpose of allowing students' learning, how they affect the meaning of knowledge, and how the various subsystems interact within a given teaching situation" (p. 143).

Research since 1990 has explored not only teaching and teachers' knowledge, with a focus on how they impact students' mathematical learning, but also how teachers come to know what they know and how their practice develops to promote students' learning. The importance of this as a research field was recognized by ICMI in the fifteenth specially commissioned ICMI study, *The Professional Education and Development of Teachers of Mathematics.* The invited international program committee (IPC) met in Prague in 2003, from which meeting a discussion document and call for papers was produced (ICMI, 2004), and worked toward an international conference in Águas de Lindóia in Brazil in 2005. From the papers submitted, 125 were accepted for work at the conference (http://stwww.weizmann.ac.il/G-math/ICMI/log_in.html). In response to recognition that publications in the field are underrepresentative of activities in many countries, a strong effort was made to include all conference participants who wished to be part of writing the study volume (Even & Ball, 2009). The volume includes two major sections reflecting two strands of activity at the conference: Strand 1, "Initial Mathematics Teacher Education," and Strand 2, "Learning in and through Practice." The chapters in the second strand relate particularly to the focus of this book. There are four chapters focusing on (1) factors, benchmarks, and issues in mathematics teacher development; (2) teacher learning as participation in social processes; (3) tools and settings supporting mathematics teachers' learning; and (4) the balance of teacher knowledge in mathematics and pedagogy. In the writing of this strand (in English), editors worked hard to be inclusive of the wide spread of authors and papers. This inclusiveness was not easy to achieve, not least because of the difficulties of communication with a large number of people electronically. Readers will judge the extent to which the goals of inclusivity were achieved.

Handbook of Mathematics Teacher Education

Further recognition of the maturing of our subfield is demonstrated in the first *Handbook of Mathematics Teacher Education,* commissioned by Sense publishers and published in 2008 (Wood, Jaworski, Krainer, Tirosh and Sullivan, 2008). The editors agreed on four areas of research and scholarship to be represented in the handbook and invited contributions from international scholars. The four sets of papers have formed volumes of the handbook as listed below.

- *Knowledge and Beliefs in Mathematics Teaching and Teaching Development* (Sullivan & Wood, 2008)
- *Tools and Processes in Mathematics Teacher Education* (Tirosh & Wood, 2008)
- *Participants in Mathematics Teacher Education: Individuals, Teams, Communities, and Networks* (Krainer & Wood, 2008)
- *The Mathematics Teacher Educator as a Developing Professional* (Jaworski & Wood, 2008)

Themes and issues permeating these volumes are taken up below.

Section Summary

This brings me to the end of an international historical perspective on our field. We see here the growth of mathematics teacher education as a subfield of mathematics education over almost three decades. The main points that I take from this, which influence strongly the themes and issues that follow, are these:

1. During the 1970s and early 1980s, the teacher was seen merely as a facilitator providing for students' learning of mathematics in an instrumental fashion interpreted in essentially simplistic terms often with connotations of deficiency.
2. A major shift toward recognizing context and complexity in teaching and the roles of a mathematics teacher was associated with (a) a shift in theoretical perspectives toward constructivism and social interactionism generally, which are seen to impact how teacher education initiatives are theorized and practically conceived, and (b) a growth in qualitative research methods, particularly case studies, which allow a more in-depth study of teaching and associated treatment of the professional development of teachers.
3. Concerns have arisen about the polarized perspectives in mathematics teacher education arising from the language of publication and conference discourse (English) and the domination of some parts of the world over what is considered worthy of study.

Considerable work is ongoing both in the conferences that continue to happen every year, two years, or four years and in special initiatives internationally. Despite point 3 above, we see more examples of international perspectives permeating the dominant language and discourse. This book constitutes one of those initiatives. I continue now to address what I see as being important themes in the education of practising teachers of mathematics. Although I strive to keep an eye on the international scene and associated contexts, my main focus is on addressing complexity in MTE and how it plays out in studies of practising mathematics teachers, their learning, and their development.

Themes in Mathematics Teaching Development

Education of Prospective and Practising Teachers

The research literature relevant to this chapter might be divided into papers on teachers and teaching, papers on educating prospective teachers, and papers on educating practising teachers. For a historical dimension, I have drawn extensively on the four review papers mentioned above, which are, chronologically, Hoyles (1992), Ponte (1994), Llinares and Krainer (2006), and Ponte and Chapman (2006). Consistent with the time at which they were published, the first two focused largely on teachers and teaching, whereas the latter two focused more on teacher education. Llinares and Krainer separate into two sections their considerations of the education of prospective and practising teachers, whereas Ponte and Chapman discuss them together, indicating in most cases which type (prospective or practising) is being discussed.

These distinctions raise the question of what we mean by programs for educating teachers. It seems clear, where prospective teachers are concerned, that many countries have a formal program (with a variety of names, including pre-service or initial teacher education programs) through which new teachers are educated and brought into the profession. In the case of practising teachers, terminology includes in-service education and continuing professional development (CPD). The former often refers to formal programs, whereas the latter refers to a range of activities. It is on research in these two areas that the following sections focus.

The Practice of Practising Teachers

As I have indicated in relation to development within PME, the first papers to deal explicitly with our subfield portrayed the teacher as deficient, reporting teacher education activity aimed at changing teaching practice in accordance with curriculum needs. A developmental axis can be seen at this time with a shift from cognitive studies focusing on the individual learner, on (largely quantitative) studies showing teacher deficiency with a gap between belief and practice, on the teacher as a facilitator or passive conveyor of knowledge, toward more qualitative (case) studies from theoretical perspectives of constructivism or social interactionism emphasizing the importance of context and complexity, a social dimension with situated beliefs, and "sensible" behaviour (Hoyles, 1992, p. 3: 275).

By the time of PME 30 and the associated review of research, there was considerable development but some disappointing conclusions, as reflected in the review papers from the PME 30 compilation (Gutiérrez & Boero, 2006). The two review papers are distinguished by their content, Ponte and Chapman (2006) focusing on teachers' knowledge and Llinares and Krainer (2006) on teachers (and teacher educators) as learners. Teachers' knowledge has become a hot topic in recent years. It is clear that teachers employ considerable knowledge in their practice of teaching and that it would be valuable to know more about the nature of this knowledge in order to inform teacher education programs. Ponte and Chapman review the literature to date in this area, focusing on the following.

- *Teachers' mathematical knowledge:* studies of the teaching of arithmetic, fractions, rational numbers, geometry, functions, word problems, reasoning, and limits. In many cases, such studies drew attention to teachers' perceived inadequacies in these topics.
- *Teachers' knowledge of mathematics teaching:* studies focusing on teachers' awareness of pupils' misconceptions; their pedagogical content knowledge; and connectedness in the teaching of mathematical concepts.
- *Teachers' beliefs and conceptions:* studies relating beliefs and knowledge; relating beliefs to particular aspects of teaching and learning, such as problem solving, students' errors, technology, and gender differences; and relating belief to practice.
- *Teachers' practices:* studies based on psychological or sociological frameworks; teachers' biographical studies; curriculum-related studies; and studies related to innovation and reform.

These focuses chart a progression from defining teachers in terms of their facilitative activity in supporting students' engagement with mathematics, through exploration of teachers' own conceptions (knowledge and beliefs) in their engagement in practice, to more deeply rooted studies, psychological or sociological, of the practice of teaching in its complexity. Ponte and Chapman quote Even and Schwartz (2002) in saying that "practice is too complex to be understood by only one perspective" and point to "the significant growth of research in mathematics teachers' practice" as "the most salient aspect of research concerning the activity of the teacher in recent years" (p. 483). From the wide range of papers reviewed, Ponte and Chapman suggest that "the most common conclusion is that teachers need further learning to carry out 'better' practices, *more aligned with the researchers' espoused perspectives*" (p. 485; emphasis added). They offer one disappointing conclusion, that still "the emergent image of the teacher is that of a professional with deficient knowledge." They recognize also a research emphasis on the complexity of teachers' knowledge and its "intimate relation" to practice (p. 486). A challenge for research is to find methodologies to explore this relationship without defining the teacher as deficient according to theoretical perspectives espoused by the researcher. This challenge is addressed further below.

Llinares and Krainer (2006, p. 439-443) focus on studies relating to teachers' learning, mainly "as a consequence of having participated in some kind of programme or course." For example, programs focus on

- *raising teachers' awareness of mathematical process and content:* with a common assumption that "for teachers to become competent in mathematics it was necessary for them to learn mathematics in the same way as they were expected to teach it";
- *raising teachers' awareness of children's mathematical thinking:* with a suggestion that "to provide opportunities for teachers to explore children's ... activity contributed to teachers' ongoing professional development"; and
- *emphasizing reflection, collaboration, and community building* in promoting teacher learning and teaching development.

Such programs have "content" goals (see Simon, 2008, discussed below) that arise from how teacher educators perceive the learning needs of teachers. Indeed, research on teaching development is largely conducted by the teacher educators who design such programs. We can see the "content" to include both mathematics per se and ways in which the learning of mathematics can be approached, for example in problem-solving environments or in students' classroom activity as a basis for learning through analysis of their ways of interacting with mathematics. Focuses on reflection, collaboration, and community building can be seen in how content is structured and activity designed for teachers' learning in these programs. Llinares and Krainer recognize the centrality of the teacher educator in such programs and ask about the ways in which teacher educators themselves learn, looking at models of learning related to those seen in programs for teachers. However, they also point out important differences in the constraints faced by each group, mainly in terms of working contexts: teachers deciding what to teach face limitations from, for example, curriculum and educational level, whereas teacher educators have more freedom in deciding the content of their work with teachers. This relative freedom extends also to teacher educators defining their own ways in which to grow professionally. I take up these ideas further below. The authors point to a fusion between teacher education and research in intervention programs in which interactions between teachers and teacher educators are a focus of study in considerations of the learning of both groups. They point out that, though such studies provide deep insights into practice through continuous interaction and communication with practice, leading to "interesting research-oriented stories" (p. 451), they are nevertheless local stories and that some means is needed to go beyond such stories to larger-scale perspectives on teachers' learning and practice.

From these reviews of papers presented at PME conferences, we see research that reveals a shift in conceptualizing teachers' thinking and practice in terms of a deeper awareness of complexity and studies that look in detail at teaching practice and teachers' knowledge and learning. Nevertheless, and disappointingly, we still see an emphasis on the teacher as *deficient* according to researchers' (teacher educators') expectations. This raises important issues that relate to the expectations of teacher educators for teaching in schools and why teaching is seen not to fulfill such expectations. The wider socio-systemic factors that contribute to activity in schools have to be recognized and addressed. Relationships between teachers and teacher educators in developing or improving teaching, the sources of knowledge influencing such relationships, and a shift from local stories to more general concepts all need to be addressed.

The Nature of the Education of Practising Teachers

So how are teacher education programs for practising teachers motivated and conceived? How is it decided what is needed and how needs can be addressed and by whom? Writing recently, Ruhama Even (2008, p. 59) refers to "the ill-defined nature of the field of educating practicing teachers." She comments on the diversity of terms for educators who work with practising teachers: common terms are "professional development (PD) providers," "professional development teachers," "professional developers," "teacher developers," "facilitators," "teacher-leaders," "teachers of teachers," "teacher educators," and "in-service teacher educators." She maintains further, with reference to Wilson and Berne (1999, p. 197), that the system of providing professional development opportunities for practising teachers is a "random, sometimes voluntary, sometimes mandated, always fragmented system" (cited in Even, 2008, p. 60).

My own experience of involvement in what has often been called in-service teacher education in the United Kingdom accords strongly with this evaluation. In 2002, I was part of a team of educators who sought to write a historical account of teacher in-service education in the United Kingdom to date. Although we knew about a variety of initiatives, many of them highly regarded in terms of their developmental outcomes, we found little or no associated published research (so such programs and their outcomes do not feature in the reviews referenced above). It seemed that, historically in the United Kingdom, professional development initiatives had not commonly had associated research programs. Where we did find research, it was mainly in the form of some kind of evaluation of a program (Macnamara, Jaworski, Rowland, Hodgen, & Prestage, 2002). It would be interesting to know the extent to which this pattern mirrors practice elsewhere. The experience certainly seems to fit with the words from Even (2008) above.

Given, then, the "ill-defined nature" of the field in providing professional development opportunities for practising teachers, what can we glean from published research and the experiences of those engaged in such provision? My own experience as a teacher educator, as *JMTE* editor, and as a reader of research studies, such as those reviewed from PME reports, suggests that programs for practising teachers include several models, which I list below. The first three see teachers progressively as *pupils, participants,* and *partners* in the educational process (Jaworski, 1999). The third and the latter two emphasize teachers' engagement in research as a developmental tool.

- Courses or summer institutes for teachers led by teacher educators: teachers as pupils (e.g., Murray, Olivier, & Human, 1999; Schifter, 1998).
- Research and/or developmental programs led by teacher educators: teachers as participants (e.g., Cobb, Wood, & Yackel, 1990; Remillard & Geist, 2002).

- Collaborative research in learning and teaching: teachers as partners in research with teacher educators/academics (e.g., Krainer, 1999; Lin, 2002).
- Research programs in which teachers explore their own practice: teachers as researchers (e.g., Jaworski, 1998; Phillips, 1997).
- Research initiatives by teachers (sometimes as part of a master's or PhD program): teachers as designers of research (e.g., Atkinson, 1994; Lee, 2006).

We might see the first model here as referring to formal education programs for teachers in which teachers are engaged with specific content, mathematical and/or pedagogical, with some attempt to enhance or change teachers' thinking and practice in a decontextualized mode. The second model might also sometimes fit this mode. On the other hand, the second through fifth models can all be seen to draw teachers into active developmental participation with educators and possibly engagement in inquiry or research that is embedded in the context and complexity of practice. We might ask in each case how teachers' learning and development relate to the nature of the program. Those responsible for designing programs need to understand, for any model, the learning taking place in relation to the experiences lived by teachers. Although there are particular examples of well-researched programs,[5] such learning experiences and their relation to the program model are not generally well documented from the point of view of research.

Programs with "Content" Goals

Martin Simon (2008, p. 18) suggests that teachers' professional development efforts can be sorted into two categories:

1. those with process goals only: the engagement of teachers in inquiry-based, reflective practices in the context of professional support and communication structures; there are no a priori learning goals for teachers involved in these programs (other than learning the processes of inquiry, reflection, etc.); and
2. those that have content and process goals: courses and workshops for teachers in which teacher educators aim to promote particular mathematical and pedagogical concepts, skills, and dispositions.

In terms of my set of models above, perhaps the first two can be seen as having both content and process goals and the last three as having process goals only. This distinction seems worth exploring further in terms of some of the questions raised above, and I start with Simon's second category: programs with both content and process goals. Llinares and Krainer (2006) have offered a range of examples of such programs. In terms of Simon's wording above, we might ask the following questions.

- What is involved in the "aim to promote"? Who promotes, how do they decide *what* to promote, and how do they *go about* promoting particular mathematical and pedagogical concepts, skills, and dispositions?
- What are the actions and roles of teachers and teacher educators in these programs, and how are they related to the learning of either group?
- What issues, tensions, and contradictions are evident?

The nature of the "content" is clearly an issue—for example, mathematical content and didactical or pedagogical content—as are the kinds of processes used to promote understanding of content. Distinctions between mathematics, didactics, and pedagogy are not clear, as the literature shows. Where mathematical knowledge is concerned,[6] teachers' knowledge of mathematics is variously referred to as

- mathematical content knowledge (Rowland, Huckstep, & Thwaites, 2005; Shulman, 1987);
- mathematical knowledge for teaching (Ball & Bass, 2003; Schifter, 1998);
- pedagogical content knowledge (Graeber & Tirosh, 2008; Shulman, 1987); and
- didactical knowledge (Bednarz & Larochelle, 1998; Brousseau, 1992; Chevallard, 1992).

Briefly, mathematical content knowledge is knowledge of mathematics per se. Mathematical knowledge for teaching goes further in the sense that the mathematical topic is known in a teaching context and in terms of how students can be expected to approach, appreciate, or have difficulty with the topic. Pedagogical content knowledge refers to the overlap between mathematics and pedagogy essential in creating a classroom environment in which students can learn mathematics. And didactical knowledge focuses on the expression of mathematical ideas in terms of material on which students will work, the tasks, examples, and situations through which mathematics can be accessible to pupils. It also refers to the organization and progression of knowledge developed by the teachers, to the criteria guiding the choice of problems and situations, to the way in which teachers exploit and interpret the solutions developed by students, and to the processes of institutionalization. To design programs in informed ways with respect to teachers' knowledge and learning, we might ask the following questions of such research:

- How do teachers come to know mathematics adequately (?) for the levels at which they teach?
- How do teachers use their mathematical knowledge to create productive (?) opportunities for students to learn mathematics?
- How do teacher educators work with teachers to promote effective (?) learning of mathematics at all levels?

The words *adequate, productive,* and *effective* are followed by a question mark to indicate their problematic nature. They are words that we use all the time and assume a common usage—of course we want learning to be effective. However, when we go below the surface and start to analyze just what each one of us means by the term "effective," we find differing ways of seeing the learning process, seeing how students (or teachers or educators) understand mathematics, which would then make a difference to what is regarded as effective.[7] For example, the much-quoted work of Skemp (1976) has distinguished between instrumental and relational understandings of mathematics. Effective instrumental understanding seems very different from effective relational understanding. This suggests that, to answer questions such as the three listed above, we have to look critically at what we mean by learning. This takes us into the realm of theory.

Theoretical Perspectives Underpinning the Designs of Programs

A discussion of theoretical perspectives on learning is beyond the scope of this chapter; however, we might distinguish a number of models of learning, each of which is linked to some well-expounded theoretical perspective. The following can be seen as relevant to current situations in classrooms.

- Direct instruction models: teachers showing, telling, demonstrating; metaphors of transmission or conveying of knowledge.

- Constructivist instructional models: the teacher as an individual "cognizer," making personal sense of what is offered in developmental programs; similarly, the student as an individual "cognizer" of mathematics.
- Reflective practice models: teachers reflecting on and in their teaching, often with an aim for development or change.
- Socio-cultural models: learning as participation; teachers learning in interaction in social settings; importance of culture and context.

Unlike the theories on which they are based, the models are rooted in practice and take on the idiosyncrasies of practice that a purely theoretical approach might question. In any style of teaching, we might recognize elements of more than one of the models with a rationalization that makes sense in context. For example, we might see programs that claim to be socio-culturally rooted, that take participation as a central construct, and that encourage reflection as a means to establish individual perspectives on the classroom. Whereas theorists might argue strongly about the epistemological precedents of such theorizing,[8] a practical rationale, based on what is done and how, can make realistic sense. However, distinguishing the models allows us to look critically at practical rationales and cut across complexities. This is true for all programs, whether focusing—as suggested by Simon (2008)—on content and process or on process only.

The distinction that Simon (2008) makes between programs that have only process goals compared with those with both content and process goals is more than categorical. It draws attention to the very nature of the teaching-learning process and the roles of those within it. Although there is a spectrum of program modes, and no well-defined format for either mode of program, if there is content to be taught, then some form of teaching or instruction might be assumed, whereas this need not be the case when no content is involved. So the ways in which learning and development are deemed to take place are presumably different in the two categories. This seems to me to be a significant issue in the aim, design, and implementation of programs, and I take it up in the rest of the chapter. A key consideration concerns the roles played by teachers and learners within these programs; indeed, who are the teachers and the learners?

Programs with Process Goals Only

So what kinds of programs fall into Simon's (2008) first and my last three categories? As examples, Simon mentions programs based on the Japanese lesson study model and programs focused on teacher inquiry or teacher research: "The basis of these programs is that the engagement of teachers in inquiry-based, reflective practices in the context of professional support and communication structures can support the ongoing professional development of mathematics teachers" (p. 18). What seems to be suggested here are programs with a professional development purpose but without "content" to be taught in any formal instructional sense.

Four papers presented at an international symposium at the American Educational Research Association (AERA) conference in 2008 can be seen to exemplify this category. The symposium focused on partnerships between teachers and academics to promote professional development in teaching, and the four papers, briefly, were as follows.

- A focus on video material recorded in a classroom in Australia. Teachers selected from this material key episodes for discussion in a research group consisting of teachers and academics. The video acted as a shared classroom experience from which participants could discuss issues. This activity provided an opportunity for teachers to modify classroom practice according to new awareness gained from the interactive work. (Gorur & Clarke, 2008)
- A focus on a nationwide bank of PD programs that was made public and financed in Germany. Teachers could elect to take part in advertised programs, choosing what they wanted to learn. Those leading the programs operated in pairs consisting of one academic and one practising teacher. Courses combined theory and practice. They were evaluated by a Centre for Pedagogical Research. (Rösken & Törner, 2008)
- A focus on collaboration between teachers and academics in China to design lessons and "upgrade" teachers' ideas. One teacher designed a first lesson and taught it, with observation from other teachers and academics. Discussion and reflection between the teacher and the observers led to a redesign of the lesson and learning for all concerned. This was the start of a design cycle involving several iterations, and it was followed by dissemination to other teachers of what had been learned in the iterative process. (Bao & Huang, 2008)
- A focus on lesson study in Japan involving a lesson study model in which teachers planned lessons for observation and analysis. Academics worked with the teachers before and after the lessons,

> which were observed by other teachers and academics. Discussion and analysis of the lessons led to new insights, issues, and awarenesses for all those involved, which in turn led to new possibilities for practice. (Shimizu, 2008)

In each case, teachers engaged with opportunities to reflect on their own practice through engagement in a PD program. A distinctive feature of the programs was their creation of partnerships between teachers and academics to promote professional development in teaching. Teachers either led the developmental process, with responses from their academic colleagues, or were partners in creating opportunities for development. Learning and development followed from critical discussion and reflection, building on practical activities in the program. Although professional development was intended for the teachers, the academics were also learners in these programs.

I have provided these examples to emphasize the distinction between programs aiming to teach content (with content and process goals) and those providing opportunities for development (with process goals only). It is clear, from research in the examples, that considerable learning took place for both teachers and academics. However, the programs did not set out to teach particular content. We might ask, what "content" was learned? To what extent can pre-identified content be addressed through such programs? These questions raise obvious issues for teacher educators designing programs, but could these be issues for teachers themselves? Might teachers themselves actually be in a better position to choose what to address in terms of content? The German program mentioned above (Rösken & Törner, 2008) offers such a possibility. Perhaps both are needed?

Teachers and Teacher Educators

As we saw above (Even, 2008), there is a range of terms used in the literature and in practice to describe the professionals who work with teachers to promote development in teaching and learning. It might be that the term used is related to the mode and nature of a particular program. However, the term "teacher educator" is commonly used to describe such professionals internationally. In the AERA symposium, the programs were referred to as partnerships between teachers and *academics.* The academics in each case were professionals from a university setting where the term "academic" would describe the nature of their job. In research projects in which I worked in Norway involving partnerships between academics and teachers, academics called themselves *didacticians.* This was a deliberate move, first to recognize our role as professionals working in the didactics of mathematics,[9] and second to avoid the term "educator" in order to avoid (if possible) the perception that our role was to instruct teachers or that teachers were not themselves educators. We wanted to create a partnership in which both partners brought knowledge and experience, and mutual respect for each other's considerable expertise, and in which both would inquire into developing mathematics teaching and learning (Jaworski, Fuglestad, Bjuland, Breiteig, Goodchild and Grevholm, 2007). Our aim was that both groups would be *co-learners* in the developmental process (Wagner, 1997).

So, in this chapter, when I use the term "teacher educator," it is (a) as a general term that is largely acknowledged in the field (some term needs to be used) and (b) with a critical view to questioning what it means, its connotations, and the diversity of experience that it encompasses. So we might start by comparing the roles of teachers and teacher educators (see Table 1.2).

By listing the points as in Table 1.2, I draw attention to the parallels in these roles. In each case, the second point emphasizes the instructional aim and the third point the didactic function of creating a learning environment. The fourth point makes clear that both teachers and educators are learners in their respective profession. This means that both generate knowledge related to their particular practice and can learn from their engagement in practice. When they work together, these forms of knowledge must be related in some ways. So, when teachers and educators work together, how is knowledge in the processes of teacher education and teaching development distributed between them? The diagram in Figure 1.1 represents one way of seeing such

Table 1.2
Comparison of roles of teachers and teacher educators

Teachers have responsibility to	**Teacher educators have responsibility to**
Work with students	Work with teachers
Promote students' learning of mathematics	Promote teachers' learning of mathematics teaching
Engage with didactics and pedagogy to create learning situations for students	Engage with didactics and pedagogy to create learning situations for teachers
Be themselves learners of mathematics teaching	Be themselves learners of mathematics teacher education

a distribution (Jaworski, 2008, p. 336). In the wider complexity of educational systems and socio-cultural settings in which teaching and learning are located (the large rectangle), the two circles represent knowledge within the two groups. The central section (B) shows that there is considerable common knowledge, albeit knowledge that is highly related to the particular professional practice and context and therefore can seem different in its manifestations. Sections A and C reflect the particular areas of knowledge related to expertise and responsibility in the two groups. It is a simple diagram that avoids many questions about knowledge addressed in the wider literature. Its main purpose is to emphasize the considerable knowledge and expertise in both groups, the professional status associated with such knowledge, and the importance of the co-learning process when the groups work together.

The nature of all these factors will depend of course on the type of program, its design, operationalization, and aim, which brings us back to some of the questions raised above, particularly in regard to the central section (B), where knowledge is shared albeit with different manifestations. It would be valuable to use research to reveal finer details of these manifestations and to explore how working together can enable each group to understand better the aims, motivations, and concerns of the other. Wagner (1997, p. 16), speaking of *co-learning agreements,* noted,

Figure 1.1
Distribution of knowledge in collaboration between teachers and educators

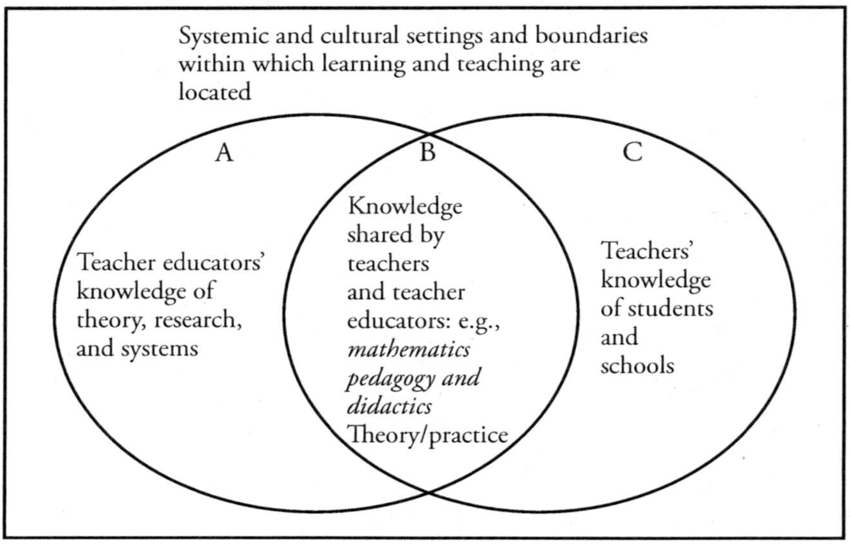

> In a co-learning agreement, researchers and practitioners are both participants in processes of education and systems of schooling. Both are engaged in action and reflection. By working together, each might learn something about the world of the other. Of equal importance, however, each may learn something more about his or her own world and its connections to institutions and schooling.

If we replace "researchers and practitioners" with "teachers and educators, both of whom are also researchers," then this quotation offers a vision of what is possible in such a partnership. Before going further with ideas of collaborative learning, however, it seems worth addressing how educators come to know what they know. How do teacher educators learn?

How Do Mathematics Teacher Educators Learn, and How Similar Is This to How Teachers Learn?

In some cases, new teachers might be accepted into schools with no formal training; however, the principal route into teaching in many countries requires new teachers to have participated and been successful in some initial or pre-service educational program (Llinares & Krainer, 2006). This is their "training" for the job, and the responsibility for this training most often rests with teacher educators, academics in university settings. Given the wide acceptance of such provision for the education of new teachers, should new teacher educators be educated similarly?

When I became a teacher educator in the United Kingdom, I had considerable experience and knowledge but no formal training. I had been a teacher of mathematics myself; I had contributed to courses for teachers at the Open University in the United Kingdom; I had engaged in my own PhD studies and research in mathematics education. In appointing me as a teacher educator, with responsibility for programs in both initial and in-service education, the university accepted these qualifications and presumably assumed that the corresponding knowledge was adequate for the job. It certainly seemed important that I had been a mathematics teacher myself, but this is not true for all academics employed as teacher educators. If educators are to work effectively with teachers who have been through an initial training program and possibly gained considerable experience in school and classroom, then in what ways can educators develop suitable knowledge that allows them to work with teachers in informed ways?

Simon (2008, p. 26) suggests that "being an effective mathematics teacher is necessary, but not sufficient, to be an effective mathematics teacher educator." He also questions whether a PhD in education is sufficient training to become a teacher educator: "In the US teacher educators are being prepared in doctoral programs without the conceptual frameworks that they require in order to work with prospective and practicing teachers. Thus the goals of their work and the developmental process that they endeavour to support and promote are under-defined." So what kinds of definition would be appropriate, and how do educators come to suitable conceptual frameworks for their work with teachers? Zaslavsky (2008, p. 95) elaborates a set of "unifying themes that reflect goals for mathematics teacher education" that "concern qualities and kinds of competence and knowledge that mathematics teacher education seeks to promote in prospective and practicing teachers in a broad sense." The themes are

- developing adaptability;
- fostering awareness of similarities and differences;
- coping with conflicts, dilemmas, and problem situations;
- learning from the study of practice;
- selecting and using (appropriate) tools and resources for teaching;
- identifying and overcoming barriers to students' learning; and
- sharing and revealing self-, peer, and student dispositions.

Looking back on my own experience, I can recognize all of these qualities and competencies as being part of my knowledge growth and development. They have come through a variety of activities alongside my professional practice—for example, reading, writing, attending seminars and conferences, undertaking research in classrooms, talking with teachers, supervising research students, and having good, critical colleagues. I suspect that many colleagues around the world have developed as teacher educators similarly. What strikes me particularly as I review this list is that the set of necessary qualities or competencies applies both to teachers and to teacher educators. So how do educators learn or develop these qualities, and is there a parallel to how teachers learn and develop?

To address this question, we might ask, what would or could it look like to design a program to educate teacher educators? Two such programs are the MEd program at the Institute for Educational Development at the Aga Khan University in Karachi, Pakistan (Farah & Jaworski, 2006), and the MANOR program at the Weizmann Institute in Israel (Even 2008). Both programs address Zaslavsky's (2008) list of competencies, both have formal instructional elements, and both encourage engagement in professional activities, including reading, writing, attending seminars and conferences, acting as critical friends, undertaking research in classrooms, and working with teachers and pupils. This list of activities is similar to the one that I gave to describe my own learning as a teacher educator, albeit not as part of a professional program. What these two programs recognize is that there are strategic activities that are key to developing as a professional educator that include reading the academic and professional literature, writing scholarly and reflective accounts relating to theory and practice, and engaging in research on mathematics learning and teaching. Although not all educators have taken part in such a professional program, most have engaged in such activity and recognize the importance of learning through such activity. Can this be the same for teachers? What different aspects of context might affect this possibility?

One of the paradoxes of teacher education initiatives is that, despite recognition of how we learn as teacher educators, we subscribe to creating very different learning situations for teachers to which teachers also subscribe. This is captured in the words of Sandy Dawson (1999, p. 148):

> This manifestation of in-service culture seems to have the following basic principle: there is something wrong with mathematics teaching worldwide and we, as mathematics educators, must fix it. Many mathematics teachers have bought into this culture. Such teachers seem to be seeking new ways to fix their practice. But this places mathematics teachers in a relationship of co-dependence with mathematics teacher educators. Mathematics teachers need someone to fix them, and mathematics teacher educators need someone to fix. The two groups seem made for each other.

As editor of *JMTE,* I read many submitted papers that seemed to subscribe to this perspective. The programs addressed were largely ones that Simon (2008) would categorize as having both content and process goals and often involved teacher educators as instructors, with teachers as pupils rather than either participants or partners. An assumption was that the expert teacher educators would instruct the teachers in teaching knowledge and the arts of teaching.

A "New Pedagogy" in Teacher Education

Chapman (2008) reviewed a set of research papers related to studies conducted by mathematics teacher educators in North America on their own programs developed for the education of prospective teachers. Her aim was "to discuss the teacher educators' learning from researching their own practice" (p. 117). With reference to Lampert, Heaton, and Ball (1994), she writes that

> a new pedagogy of teacher education is required if teacher education is to prepare prospective teachers to be responsive to visions of mathematics education advocated by the reform. They [Lampert, Heaton, & Ball] argued that conventional teacher education programs, which present prospective teachers with ideal methods and techniques derived from a synthesis and interpretation of educational theory and research, do not represent the complexity and uncertainty of teaching found in reform-based instruction. (pp. 115-116)

Although Chapman, along with Lampert, Heaton, and Ball, refers here to pre-service programs, the words could apply as well to in-service teacher education. Chapman points out that research conducted by teacher educators on their own practices has the potential to provide professional development for educators and allow them to know something of significance about their own practice (see also Jaworski, 2003). However, her survey of relevant articles shows, for the most part, that this learning was presented in the articles as what other teacher educators could learn about the nature of these approaches to teacher education rather than as the actual learning of those conducting the research. These authors offer what is learned from research as generalizable knowledge for the scientific community or as advice for other professionals rather than as informing their own professional practice.

Thus, what has been found here is that teacher educators' learning from their own research on their own programs is not usually cast as teacher educators' learning in practice, though it manifestly is such. Several *JMTE* papers stand out for me as making this link clearer: authors synthesize their findings for a general academic presentation (as encouraged by a quality research journal) and highlight the importance of their findings for their own thinking and practice (see, e.g., Goodell, 2006; Heaton & Mickelson, 2002; McDuffie, 2004). What is clear from a wide range of papers in *JMTE* is that there has been a significant shift from programs in which teacher educators instruct teachers according to teacher educators' own perceptions of how to know and how to teach toward a more thoughtfully critical conception of teacher education. Simon (2008) marks this as a contrast between *perception-based* teaching and *conception-based* teaching. In the latter, the teacher seeks to become more aware of the conceptions of the students in order to construct relevant learning experiences rather than pursue mainly the teacher's own perceptions of how things should be (see also Tzur, 2008). It seems clear, and Simon makes this point, that these notions apply similarly to teachers teaching students and educators teaching teachers. Paola Sztajn (2008) takes this further in a discussion of the idea of *reciprocity* in *caring* relationships. When the teacher is seen by the student overtly to enter into students' ways of thinking and to work with students' conceptions, the student is more able to appreciate what the teacher can offer from his or her own knowledge and experience.

From my own experience in professional practice, in international meetings, and as editor of *JMTE,* I detect a shift in perspective in the mathematics education community in addressing teacher education. I wrote about this as follows, referring to current writing:

> [There has been] a shift in tone and nuance in the ways educators write about educating teachers. There is less of a surety of models of practice that educators promote with teachers and much more a sense of uncertainty. With this uncertainty comes, almost paradoxically, a strength of purpose, new ways of speaking about mathematics teacher education, and new paradigms of practice. These build on notions of reflection, for both teachers and teacher educators, on teacher-as-researcher and simultaneously educator-as-researcher, and on growing recognitions of epistemology, of complexity and the importance of not trying to oversimplify. (Jaworski, 2008, p. 338)

This quotation comes from the fourth volume of the *Handbook of Mathematics Teacher Education* (Jaworski & Wood, 2008), entitled *The Mathematics Teacher Educator as a Developing Professional.* Perhaps the fact that we have a volume of a handbook focusing on the mathematics teacher educator is itself one source of evidence for what I claim. One important piece of evidence, in my view, is the increasing number of programs reported that seek to build partnerships between educators and teachers and study the processes and outcomes.

Partnerships in Professional Practice

We can see both teachers and teacher educators as learners in practice, reflecting critically on activity, possibly by engaging in research. Both bring specialist knowledge and expertise to the partnership. Both have specialist roles and responsibilities in the joint practice (e.g., Jaworski, 2001; Krainer, 2008).

The four programs reported above from the AERA symposium are examples of such partnerships, as they claim overtly. Many of Konrad Krainer's writings reflect partnerships through which Krainer (1998) has conceptualized his two-dimensional model of Action, Reflection, Autonomy, and Networking. In one dimension, we see action ←→ reflection not as polarization but as recognition of their complementary importance for learning and development. Similarly for autonomy ←→ networking: professionals are both thinking individuals and members of communities of practice in which collaboration supports learning. So one axis concerns engagement, and the other concerns the professional landscape in which engagement is rooted. From my own experience of such partnerships over many years, I suggest the following characteristics of partnerships.

- Partnerships avoid a hierarchy and seek mutual respect and reciprocity—each seeking to fulfill the goals of the other as well as its own goals (e.g., Burton, 1999).
- Partnerships assume a shared power base in which decisions on what is needed and how to fulfill needs are taken jointly through discussion and negotiation (e.g., Farah & Jaworski, 2006).
- Issues and tensions lead to learning and development (e.g., Jaworski & Goodchild, 2006).

At this point, I am ready to take up some of the questions highlighted above and to discuss some of the associated issues for the development of mathematics teaching.

Issues Relating to the Education of Practising Teachers

This book is about the professional development of practising teachers. This involves questions relating to how teachers learn and the kinds of programs that facilitate their learning. It involves considerations of what teachers know, how this knowledge relates to their teaching or is adequate for their teaching, and how this knowledge develops, particularly in relation to practice in the learning and teaching of mathematics. In the above sections, I addressed themes relating to these considerations. Particularly, I have moved toward considerations of the roles and development of those who teach teachers, the teacher educators, and how their learning and development both parallel and influence that of teachers. In this final section, I recognize issues that arise for research in the teacher education community (including teachers and teacher educators) as it seeks to improve professional development opportunities and concomitantly the experiences widely offered to learners of mathematics.

Theoretical Basis of Research Programs Reporting Teacher Education Outcomes

The move toward a new pedagogy suggests that teachers are not seen as *pupils* in professional programs; they are either *participants* or *partners.* But what do we mean by these terms? How is the nature of participation defined? Well, perhaps a starting point is to avoid seeing teacher knowledge as received wisdom (as in direct instructional models of teacher education) and to conceptualize it either in social constructivist or socio-cultural terms. This theoretical conceptualizing influences the models of practice that emerge. Social constructivist models emphasize social interaction as promoting personal reflection and leading to new insights for individuals in relation to their practices as teachers or educators (e.g., Tzur, 2008). Socio-cultural models emphasize ways of being and thinking in relation to communities of learners and teachers and the institutional settings in which they participate (e.g., Goos, 2008). In my experience, many research papers submitted to a journal, focusing on teacher education programs, are not clear enough about the underlying theories of learning and teaching that influence the programs studied. Better clarity seems to be imperative in understanding the basis of a program and research findings from a study of it. For example, Tzur (2008) makes clear that his research is conducted from a largely cognitive perspective in which conception-based learning is theorized with reference to von Glasersfeld's constructivism and rooted in Piaget and Dewey. Goos (2008) foregrounds the social, embedding her research theoretically in Vygotskian theory and Valsiner's zone theory of child development. With such clarity on theoretical perspectives, it is possible for the reader to judge reported outcomes in relation to the perspectives that have informed interpretation and analysis.

A further consideration lies in the relationships between the theoretical basis of research and its compatibility with the theoretical basis of the program studied; this also requires consideration of how research and development are linked to each other. I address this through consideration of the related roles of those involved in a program.

Collaboration between Teachers and Educators: The Nature and Complexity of Relationships

As indicated in references to Chapman (2008) above, teacher educators as researchers, reporting from their research programs, often seem to objectify their findings and isolate them from the developmental learning experiences to which they relate. My own experience suggests that most research programs on the learning and practice of teachers, in which researchers themselves are teacher educators, are related fundamentally to professional learning both for the participants and for the researchers (Jaworski, 2003). By *participants* here, I refer to both teachers and educators. Thus, the educators are both central participants in the practices studied and researchers in studying them (both insider and outsider researchers; see Goodchild, 2007; Jaworski, 2004). What is reported from the research therefore needs to be acknowledged as being dependent not only on the theoretical stance taken by the researchers but also on the theoretical roots of the developmental program and relationships between the two. The roles of teachers and educators in such programs can be related fundamentally to the theoretical stance, as can the ways in which these roles are interpreted and analyzed. When teachers take on overtly an inquiry or research role in a program, relationships between roles become even more complex—both teachers and educators learn through their roles as insider researchers, and educators take on the additional role of synthesizing more generally from the research.

Thus, when we consider roles of teachers as either *participants* or *partners* in research and development, we are making theoretical statements. We theorize both the positions of teachers within the developmental program and how we observe and analyze these programs. Speaking developmentally, it seems important to juxtapose learning and experience of teachers and educators; in research terms, needed is critical scrutiny of who is making interpretations and judgments and how they relate to the positioning of both groups of participants. I accord with Cochran Smith, and Lytle (1999) and Wells (1999) in suggesting “inquiry” as a theoretical-methodological stance that allows practitioners, teachers, and educators to look critically at their own practice while engaging with it. Thus, the role of insider researcher allows a “critical alignment” with practice that allows practitioners to engage with the norms and expectations within the community of practice of which they are a part and, at the same time, look critically toward ways of developing and improving the practice (Jaworski, 2006).

Proximity to the Classroom: Outcomes for Students' Learning Mathematics

Within the complexity just mentioned is the dimension of educator and teacher learning vis-à-vis student learning of mathematics. We might envisage nested dimensions of

- a mathematics classroom with students learning mathematics;
- a mathematics teacher education program with teachers learning mathematics teaching; and
- research on mathematics teaching development with teachers and educators learning how to improve mathematics teaching and learning.

Zooming in allows us to look closely at students' learning of mathematics and issues in classroom settings. Zooming out takes us into issues of teaching and developing teaching (see Figure 1.2).

The model here might be considered alongside the one presented in Figure 1.1, considering dimensions of teacher and educator knowledge together with the related dimensions of student, teacher, and educator learning in practice.

Figure 1.2
Nested model of educator and teacher learning vis-à-vis student learning

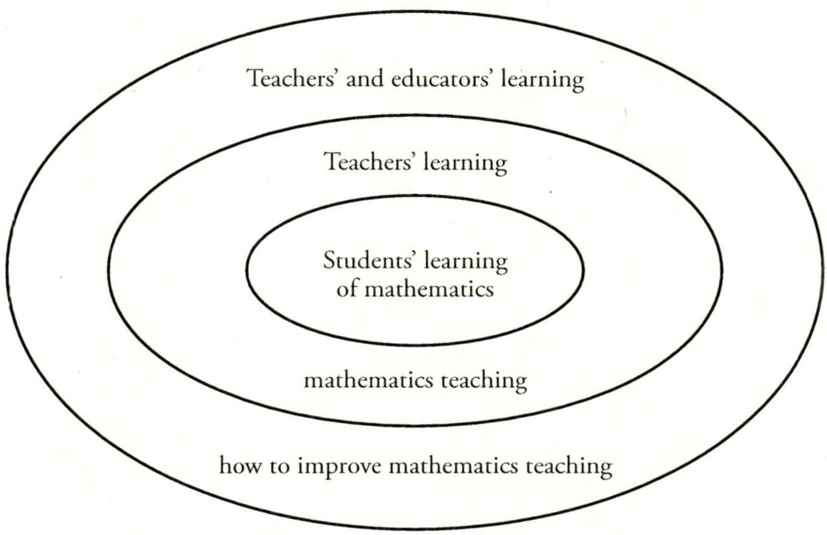

Since all teacher education is premised on the development of opportunities for students to learn mathematics effectively, their mathematical learning needs to be central to all teacher educational concerns.

When we read the literature, we see that students' learning of mathematics is implicitly or explicitly a central factor in most programs, but often both the mathematics and the students are backgrounded in conceptualizations of outcomes. It seems that, when we zoom out to express the learning of teachers and educators, the learning of students, and the mathematics that they learn, become at best implicit in what we report. This has been especially evident in my own research, and I have struggled to make sense of the complexity in a holistic sense. For example, in the Mathematics Teacher Enquiry Project (Jaworski, 1998), my focus as an *outsider* researcher was on teachers' engagement with inquiry processes as they reflected critically on their own teaching, whereas my focus as an *insider* researcher was on how my own practice developed within this program and how my actions and those of a colleague in interactions with teachers influenced the development of the program. In practice, the teacher educators (I and my colleague) were one day in the classroom with teachers and pupils, observing mathematics teaching and learning and discussing the pupils' mathematics with their teachers, on another day leading meetings of all teachers in the project and facilitating their reflections, and on another day reflecting ourselves and struggling with issues within the project. Thus, issues relating to insider and outsider roles were hard to separate in practice. When writing about them, it seemed to make sense to distinguish between the different layers, though this had the effect of fragmenting the overall complexity. Thus, what we might call the "zooming problem" alerts us to ensure (if we can) that we do not lose sight of students' learning of mathematics within the complexity of issues in teacher education.

Formality of Teacher Education Processes

As I wrote the above paragraph, I had in mind a developmental research process in which teachers and educators together explored aspects of learning and teaching and their development. Often such programs are characterized as research and/or development programs without being addressed as mainstream "teacher education." They are programs that have *process* goals rather than explicit *content* to be taught—that is, where there is something to be taught and someone to teach it (Simon, 2008). Research shows, however, that considerable learning happens through such programs, and as discussed above we can point to "content" in what is learned. For example, in the program mentioned above, I learned about tensions between wanting certain outcomes from practice and not knowing how best to work with the teacher to achieve them.

It might be that a developmental research program highlights learning needs that a more formal program can then address. More often, perhaps, perceived deficiencies in teaching practice (Dawson, 1999) are the motivation for more formal content-based programs. So we recognize differences between formal (content-based) teacher education programs and less formal, more developmental programs. I am not suggesting that content-based programs are not needed—we can find many justifications for them in the literature and in practice, and teachers themselves can identify needs for such provision (Rösken & Törner, 2008). However, it can seem that the outcomes of content-based programs are more distant from developments in practice than are the outcomes of developmental programs. In a more global characterization of mathematics teacher education, how should or can the two kinds of programs sit side by side, complementing each other as they contribute to teaching development? It is important that such questions, which go beyond the nature of specific programs or a specific type of program, be addressed at a socio-systemic or -political level in defining mathematics teacher education.

Degree of Autonomy of the Teacher

Central to such questions, in all of the programs considered, is the position of the teacher and the ultimate mathematics learning of his or her pupils. I have suggested a number of ways of seeing teachers in education programs (pupils, participants, partners, researchers). However, for sustained impact on education more broadly, should it be left to the individual program to decide, implicitly or explicitly, how teachers are treated? It seems fair to say that teacher education programs have moved away from direct instruction models and toward more inclusion of teachers in conceptualization and decision making. I cannot be so confident, however, about the political decision making that commissions and resources programs. It can seem that politicians and those who manage education nationally and internationally work from visions grounded in absolutist epistemological frames (Ernest, 1991) and look for outcomes in terms of measurable evidence and warrants (Shavelson, Phillips, Towne, & Feuer, 2003). One result is that the programs that find favour and support are of the content-based, direct-instructional type in which measuring outcomes seems easier than in developmental models.

Despite these remarks, there is evidence in some parts of the world that national programs move toward more complex models in which critical teacher centrality is acknowledged. This book provides examples from Brazil in which teachers are included centrally in developmental programs (see, e.g., the chapters by Fiorentini et al.; Freitas & Fiorentini; Passos & Lamonato). We can also see this happening in Austria in the IMST program of which Krainer (2008) writes. In the United Kingdom, we have had a long period of political control of education in which measures have predominated through national tests and government curricular control (Alexander & Flutter, 2009). Now we see, encouragingly, more inclusion of teachers in national policies relating to education and in decision making in teacher education programs (http://www.ncetm.org.uk). Where mathematics is concerned, there is acknowledgement of mathematics in the curriculum as needing special consideration and new vision—a "radical culture change" in continuous professional development is suggested (Smith 2004, p. 111). Other chapters in this book suggest similar trends in other nations. It therefore seems that the climate of understanding between educators and educational managers is shifting toward more open discussion about how we conceive of teacher education in mathematics globally.

Teachers and Educators as Part of a Culture Change

A climate of understanding in which teachers are included in decision-making processes about educational development brings teachers into a more central position, both of opportunity and of responsibility. As teachers are included centrally in research programs and consulted on developmental issues, they face new demands on their thinking and practice. In socio-cultural terms, we might describe the growth of teachers in new areas of opportunity and responsibility as a *culture change*. Although culture change can seem essential to effective development, it can simply shift the axes without changing practice fundamentally.

What are the implications for mathematics learning and teaching? We might see two fundamentals of mathematics education to involve, for teachers, a deep understanding of the mathematics that they teach and a mature awareness of students' engagement with mathematics and issues in fostering their understanding. Educators have a responsibility to work with teachers on such fundamentals; their own practice, however, is no more clearly defined than is that of teachers. They too work within the culture and are defined relative to social and political forces. They have to respond to both schools and teachers and to the educational managers and politicians who resource programs. I refer here to what is expressed in Figure 1.1 as "systemic and cultural settings and boundaries within which learning and teaching are located." It is the containing area in which teacher and educator knowledge, practice, and interaction are located. Although individual programs can, to some extent, set their own contextual boundaries, the more global context consists of both the sum of its educational parts and its own socio-political frame embedded in global trends and positioning. Krainer (2001, 2008) addresses the institutional demands on mathematics teacher education in relation to the spectrum of fundamentals that concerns not only teachers and educators but also "economics experts, educational policy makers, mathematicians, parents, etc." (2008, p. 194). This is one approach to a more global frame.

Possibilities: The Macro and the Micro

Lerman (1998) uses the metaphor of the zooming of a lens to discuss alternative perspectives in differing degrees of focus. The complexity of our subfield lies in the need to see at the same time both the macro and the micro—the global picture that I have referred to above in contrast to a focus on the student in the classroom engaging with (struggling with?) mathematics (the central part of Figure 1.2). I know from experience that it is hard to address both perspectives together. It can seem that research focuses finely on classroom interactions and fine-grain details of pupil or teacher learning, ignoring the wider socio-systemic influences, or it can take a more socio-systemic perspective but give little insight into finer details of practice. With my colleague Despina Potari, I have tried recently to bridge the macro/micro gap in a study of teaching using an activity theory perspective (Jaworski & Potari, 2010). With Simon Goodchild, I have also used an activity theory perspective to trace teaching-learning issues across a large developmental project, identifying tensions in practice that illuminate several levels or layers of culture (Jaworski & Goodchild, 2006). Inevitably, such papers leave the authors open to criticisms of insufficiently detailed treatments that allow depth of insight. We need more debate in the research community on how to juxtapose a range of zooms on practice to present more holistic perspectives on teaching and teaching development.

Toward an Agenda for the Future

I offer a number of points as a brief synthesis from themes and issues above to suggest a forward-looking agenda for the education of practising teachers of mathematics.

1. Teachers should be included as partners in teacher education initiatives with their teacher educator colleagues, with mutual respect for the knowledge and expertise of both partners and complementary contributions to the nature and design of programs.
2. Programs with content goals should be linked closely to developmental programs so that content can be related clearly to desired learning outcomes made evident through collaborative process-based activity.
3. The combination of (1) and (2) above should allow for exploration of knowledge in teaching without defining the teacher as deficient.
4. There needs to be recognition that teacher educators are also learners and that developmental programs provide an environment for their learning just as much as they do for teachers.
5. Development needs to be closely linked to research. Research not only studies the progress of development but also provides an inquiry base for the development of practice.
6. Studies of mathematics teaching and its development need to be designed to address complexity so that it is possible to keep in sight students' learning of mathematics while exploring broader aspects of teacher and educator development.
7. The theoretical perspectives underpinning research and development need to be made explicit so that the principles on which programs are founded and on which development is premised are clear.
8. Attention needs to be paid overtly to language issues in reporting research and dominance of the English-speaking world in research agendas. Editors of the major journals and organizers of international events have to be urged to seek new ways of promoting more equitable international communication.

These points are easy to write, and in the writing they seem like common sense. However, each one is challenging and problematic to achieve. We therefore need research in the subfield that addresses how we go about this agenda and what we learn from engaging in studies that seek to achieve these aims.

Acknowledgement

I would like to thank Nadine Bednarz warmly for her very helpful comments on earlier drafts of this chapter.

Notes

1 A personal note is perhaps also in order. I came into university education in 1984 after a career in school teaching. In my final five years of school teaching, I was head of a mathematics faculty in a large secondary school teaching mathematics to the pre-university level. I took it as a challenge to look critically at how we taught mathematics in the school and to consider approaches based on investigations and open-ended problems. We held many meetings, seminars, and workshops in the school, inviting colleagues from the academic world to help us think through our possibilities and associated issues. When I was offered the opportunity to move into university education and engage in research, it was with the motivation of investigating mathematics teaching further, of gaining insight into new ways of creating environments for students' mathematical learning. From that time, as a teacher educator and researcher, I had the privilege to work with some outstanding teachers who helped me to work on my visions with a firm base in the reality of schools and classrooms. My book *Investigating Mathematics Teaching: A Constructivist Enquiry* was the outcome of my first years of research in this field (Jaworski, 1994). Since then, I have been concerned with how mathematics teaching can develop, focusing largely on teachers' classroom inquiry and on partnerships between teachers and academics in researching teaching. I was also editor of the *Journal of Mathematics Teacher Education* for six years. I have therefore been personally close to the development of and evolution in mathematics teacher education for more than twenty years.

2 In the field of mathematics education, we might see the annual conference of PME, the international group for the Psychology of Mathematics Education, as being the major meeting ground and a principal source of publication. Although PME papers are short, and often cannot do justice to complex areas of research, they are markers in the field. The rigorous review process that has developed over the years and the annual occurrence of the conferences have meant that we have gained a reliable record of research development over more than thirty years.

3 For the first three years, three issues a year were published, increasing to four issues in 2001. At the end of 2001, editorship passed to me (I was then at the University of Oxford), and I invited an editorial team to join me (Terry Wood, Purdue University, United States; Konrad Krainer, University of Klagenfurt, Austria; and Peter Sullivan, Monash University, Australia). We continued with four issues a year until 2005. In the meantime, Kluwer had been taken over by Springer, and the Springer editor, along with the *JMTE* team, was keen to see the journal expand. Thus, in 2006, with a new member of the editorial team (Dina Tirosh, University of Tel Aviv, Israel), *JMTE* moved to six issues per year. In 2008, Peter Sullivan took over as editor in chief.

4 See http://www.springer.com/education+%26+language/mathematics+education/journal/10857.

5 We might point particularly to the project in Cognitively Guided Instruction (e.g., Fennema et al., 1996) and the classroom experiments of Cobb, Wood, and Yackel (1990).

6 A special edition of *For the Learning of Mathematics* (in production) will focus on this question.

7 See Brophy (1986) and Confrey (1986) for a debate on the "teacher effectiveness" research.

8 I am thinking of some of the big debates over the years, such as Brophy-Confrey or Lerman-Steffe (Brophy, 1986; Confrey, 1986; Lerman, 1996; Steffe & Thompson, 2000).

9 The mathematics education department is called, in Norwegian, Matematikk Didaktikk.

References

Adler, J., Ball, D., Krainer, K., Lin, F. L. & Novotna, J. (2005). Reflections on an emerging field: Researching mathematics teacher education. *Educational Studies in Mathematics, 60,* 359-381.

Adler, J., & Jaworski, B. (2009). Public writing in the field of mathematics teacher education. In R. Even & D. L. Ball (Eds.), *The professional education and development of teachers of mathematics* (pp. 249-254). New York: Springer.

Alexander, R. J., & Flutter, J. (2009). *Towards a new primary curriculum: A report from the Cambridge Primary Review. Part 1: Past and present.* Cambridge, UK: University of Cambridge, Faculty of Education.

Association of Teachers of Mathematics (ATM). (1987). *Teacher is (as) researcher.* Derby, UK: Author.

Atkinson, S. (1994). Rethinking the principles and practice of action research: The tensions for the teacher-researcher. *Educational Action Research, 2*(3), 383-402.

Balacheff, N. (1990). Future perspectives for research in the psychology of mathematics education. In P. Nesher & J. Kilpatrick (Eds.), *Mathematics and cognition* (pp. 135-148). Cambridge, UK: Cambridge University Press.

Ball, D. L., & Bass, H. (2003). Towards a practice-based theory of mathematical knowledge for teaching. In B. Davis & E. Simmt (Eds.), *Proceedings of the 2002 annual meeting of the Canadian Mathematics Education Study Group* (pp. 3-14). Edmonton: Canadian Mathematics Education Study Group.

Bao, J., & Huang, R. (2008). Teacher professional development in China: Keli (exemplary lesson development). Paper presented at the Interactive Symposium, Teacher Academic Partnerships: International Approaches to Teacher Professional Development, Division K, Section 8, Conference of the American Educational Research Association (AERA), New York.

Bednarz, N., & Larochelle, M. (1998). *Constructivism and education.* Cambridge, UK: Cambridge University Press.

Brophy, J. (1986). Teaching and learning mathematics: Where research should be going. *Journal for Research in Mathematics Education, 17*(5), 323-346.

Brousseau, G. (1992). Didactique: What it can do for the teacher. In R. Douady & A. Mercier (Eds.), *Research in didactique of mathematics* (pp. 7-40). Paris: La Pensée Sauvage.

Burton, L. (Ed.) (1999). *Learning mathematics: From hierarchies to networks.* London: Falmer Press.

Chapman, O. (2008). Mathematics teacher educators' learning from research on their instructional practices: A cognitive perspective. In B. Jaworski & T. Wood (Eds.), *The mathematics teacher educator as a developing professional,* vol. 4 of *International handbook of mathematics teacher education* (pp. 115-134). Rotterdam: Sense Publishers.

Chevallard, Y. (1992). Fundamental concepts in didactics: Perspectives provided by an anthropological approach. In R. Douady & A. Mercier (Eds.), *Research in didactique of mathematics* (pp. 131-167). Paris: La Pensée Sauvage.

Cobb, P., Wood, T., & Yackel, E. (1990). Classrooms as learning environments for teachers and researchers. In R. B. Davis, C. A. Maher, and N. Noddings (Eds.), *Constructivist views of the teaching and learning of mathematics, Journal for Research in Mathematics Education, monograph number 4* (pp. 125-146). Reston, VA: National Council of Teachers of Mathematics.

Cochran Smith, M., & Lytle, S. L. (1999). Relationships of knowledge and practice: Teacher learning in communities. *Review of Research in Education, 24,* 249-305.

Confrey, J. (1986). The critique of teacher effectiveness research in mathematics education. *Journal for Research in Mathematics Education, 17*(5), 347-360.

Confrey, J., & Kazak, S. (2006). A thirty-year reflection on constructivism in mathematics education in PME. In A. Gutiérrez & P. Boero (Eds.), *Handbook of research on the psychology of mathematics education: Past, present, and future* (pp. 305-346). Rotterdam: Sense Publishers.

Davis, R. B., Maher, C. A., & Noddings, N. (1990). *Constructivist views on the learning and teaching of mathematics, Journal for Research in Mathematics Education, monograph number 4.* Reston, VA: National Council of Teachers of Mathematics.

Dawson, S. (1999). The enactive perspective on teacher development: "A path laid while walking." In B. Jaworski, T. Wood, & S. Dawson (Eds.), *Mathematics teacher education: Critical international perspectives* (pp. 148-162). London: Falmer Press.

Ellerton, N. F. (1999). *Mathematics teacher development: International perspectives.* Perth: Meridien Press.

Ernest, P. (1991). *The philosophy of mathematics education.* London: Falmer Press.

Even, R. (2008). Facing the challenge of educating educators to work with practising mathematics teachers. In B. Jaworski & T. Wood (Eds.), *The mathematics teacher educator as a developing professional,* vol. 4 of *International handbook of mathematics teacher education* (pp. 57-74). Rotterdam: Sense Publishers.

Even, R., & Ball, D. L. (2009). *The professional education and development of teachers of mathematics.* New York: Springer.

Even, R., & Schwartz, B. (2002). Implications of competing interpretations of practice to research and theory in mathematics education. In A. D. Cockburn & E. Nardi (Eds.), *Proceedings of the 26th PME Conference, 2,* 337-344. Norwich, UK: University of East Anglia.

Farah, I., & Jaworski, B. (2006). *Partnerships in educational development.* Oxford: Symposium Books.

Fennema, E., Carpenter, T. P., Franke, M. L., Levi, L., Jacobs, V. R., & Empson, S. B. (1996). A longitudinal study of learning to use children's thinking in mathematics instruction. *Journal for Research in Mathematics Education, 27,* 403-434.

Goodchild, S. (2007). Inside the outside: Seeking evidence of didacticians' learning by expansion. In B. Jaworski, A. B. Fuglestad, R. Bjuland, T. Breiteig, S. Goodchild, & B. Grevholm (Eds.), *Learning communities in mathematics* (pp. 189-204). Bergen, Norway: Caspar Forlag.

Goodell, J. E. (2006). Using critical incident reflections: A self-study as a mathematics teacher educator. *Journal of Mathematics Teacher Education, 9*(3), 221-248.

Goos, M. (2008). Sociocultural perspectives on learning to teach mathematics. In B. Jaworski & T. Wood (Eds.), *The mathematics teacher educator as a developing professional,* vol. 4 of *International handbook of mathematics teacher education* (pp. 75-92). Rotterdam: Sense Publishers.

Gorur, R., & Clarke, D. (2008). "Encounter" learning rather than "embedded" learning: Using the case methods approach for professional learning. Paper presented at the Interactive Symposium, Teacher Academic Partnerships: International Approaches to Teacher Professional Development, Division K, Section 8, Conference of the American Educational Research Association (AERA), New York.

Graeber, A., & Tirosh, D. (2008). Pedagogical content knowledge: Useful concept or elusive notion? In P. Sullivan & T. Wood (Eds.), *Knowledge and beliefs in mathematics teaching and teaching development,* vol. 1 of *International handbook of mathematics teacher education* (pp. 117-132). Rotterdam: Sense Publishers.

Grevholm, B., & Ball, D. (2008). The professional formation of teachers. Report from working group 2. In M. Menghini, F. Furinghetti, L. Giacardi, & F. Arzarello (Eds.), *The first century of the International Commission on Mathematical Instruction (1908-2008): Reflecting and shaping the world of mathematics education* (pp. 265-276). Rome: Encyclopaedia Italiana.

Grouws, D. A., & Cooney, T. J. (1988). *Effective mathematics teaching.* Reston, Va: Lawrence Erlbaum Associates and NCTM.

Gutiérrez, A., & Boero, P. (2006). *Handbook of research on the psychology of mathematics education: Past, present, and future.* Rotterdam: Sense Publishers.

Heaton, R. M., & Mickelson, W. T. (2002). Learning and teaching of statistical investigation in teaching and teacher education. *Journal of Mathematics Teacher Education, 5*(1), 35-59.

Hoyles, C. (1992). Illuminations and reflections: Teachers, methodologies, and mathematics. In W. Geeslin & K. Graham (Eds.), *Proceedings of the sixteenth PME conference* (pp. 3-263–3-286). Durham: University of New Hampshire.

International Commission on Mathematical Instruction (ICMI). (2004). The fifteenth ICMI study: The professional education and development of teachers of mathematics. *Journal of Mathematics Teacher Education, 7*(3), 279-293.

Jaworski, B. (1994). *Investigating mathematics teaching: A constructivist enquiry.* London: Falmer Press.

Jaworski, B. (1998). Mathematics teacher research: Process, practice, and the development of teaching. *Journal of Mathematics Teacher Edu*cation, *1*(1), 3-31.

Jaworski, B. (1999). Mathematics teacher education research and development: The involvement of teachers. *Journal of Mathematics Teacher Education, 2*(2), 117-119.

Jaworski, B. (2001). Developing mathematics teaching: Teachers, teacher educators, and researchers as co-learners. In F.-L. Lin & T. J. Cooney, (Eds.), *Making sense of mathematics teacher education* (pp. 295-320). Dordrecht: Kluwer.

Jaworski, B. (2003). Research practice into/influencing mathematics teaching and learning development: Towards a theoretical framework based on co-learning partnerships. *Educational Studies in Mathematics, 54*(2-3), 249-282.

Jaworski, B. (2004). Insiders and outsiders in mathematics teaching development: The design and study of classroom activity. In O. Macnamara & R. Barwell (Eds.), *Research in mathematics education: Papers of the British Society for Research into Learning Mathematics,* vol. 6 (pp. 3-22). London: BSRLM.

Jaworski, B. (2006). Theory and practice in mathematics teaching development: Critical inquiry as a mode of learning in teaching. *Journal of Mathematics Teacher Education, 9*(2), 187-211.

Jaworski, B. (2008). Development of mathematics teacher educators and its relation to teaching development. In B. Jaworski & T. Wood (Eds.), *The mathematics teacher educator as a developing professional,* vol. 4 of *International handbook of mathematics teacher education* (pp. 335-361). Rotterdam: Sense Publishers.

Jaworski, B., Fuglestad, A.-B., Bjuland, R., Breiteig, T., Goodchild, S., & Grevholm, B. (2007). *Learning communities in mathematics.* Bergen, Norway: Caspar Forlag.

Jaworski, B., & Goodchild, S. (2006). Inquiry community in an activity theory frame. In J. Novotna, M. Kratka, & N. Stelikova (Eds.), *Proceedings of the 30th conference of the International Group for the Psychology of Mathematics Education,* vol. 3 (pp. 353-360). Prague: Charles University.

Jaworski, B., & Potari, D. (2010). Bridging the macro- and micro-divide: Using an activity theory model to capture sociocultural complexity in mathematics teaching and its development. *Educational Studies in Mathematics, 72,* 219-236.

Jaworski, B., & Wood, T. (2008). *The mathematics teacher educator as a developing professional.* Vol. 4 of *International handbook of mathematics teacher education.* Rotterdam: Sense Publishers.

Jaworski, B., Wood, T., & Dawson, A. J. (1999). *Mathematics teacher education: Critical international perspectives.* London: Falmer Press.

Krainer, K. (1998). Dimensions of teachers' professional practice: Action, reflection, autonomy, and networking. In T. Breiteig & G. Brekke (Eds.), *Theory into practice in mathematics education: Proceedings of the second Nordic conference (Norma 98) on mathematics education* (pp. 36-44). Kristiansand, Norway: Høgskolen i Agder.

Krainer, K. (1999). PFL-mathematics: Improving professional practice in mathematics teaching. In B. Jaworski, T. Wood, & S. Dawson (Eds.), *Mathematics teacher education: Critical international perspectives* (pp. 102-112*)*. London: Falmer Press.

Krainer, K. (2001). Teachers' growth is more than the growth of individual teachers: The case of Gisela. In F.-L. Line and T. J. Cooney (Eds), *Making sense of mathematics teacher education* (pp. 271-294). Dordrecht: Kluwer.

Krainer, K. (2008). Reflecting the development of a mathematics teacher educator and his discipline. In B. Jaworski & T. Wood (Eds.), *The mathematics teacher educator as a developing professional,* vol. 4 of *International handbook of mathematics teacher education* (pp. 177-200). Rotterdam: Sense Publishers.

Krainer, K., Goffree, F., & Berger, P. (1999). On research in mathematics teacher education. In I. Swank (Ed.), *European research in mathematics education,* vol. 3 (pp. 1-246). Osnabrück, Germany: Forchungsinstitut für Mathematikdidaktik.

Krainer, K., & Wood, T. (2008). *Participants in mathematics teacher education: Individuals, teams, communities, and networks,* vol. 3 of *International handbook of mathematics teacher education.* Rotterdam: Sense Publishers.

Lampert, M., Heaton, R., & Ball, D. L. (1994). Using technology to support a new pedagogy of mathematics teacher education. *Journal of Special Education Technology, 12*(3), 276-298.

Lee, C. (2006). *Language for learning mathematics: Assessment for learning in practice.* Buckingham: Open University Press.

Lerman, S. (1996). Intersubjectivity in mathematics learning: A challenge to the radical constructivist paradigm? *Journal for Research in Mathematics Education, 27*(2), 133-150.

Lerman, S. (1998). A moment in the zoom of a lens: Towards a discursive psychology of mathematics teaching and learning. In A. Olivier & K. Newstead (Eds.), *Proceedings of the 22nd PME conference* (pp. 1-66-1-81). Stellenbosch, South Africa: University of Stellenbosch.

Lin, F.-L., & Cooney, T. J. (2001). *Making sense of mathematics teacher education.* Dordrecht: Kluwer.

Lin, P.-J. (2002). On enhancing teachers' knowledge by constructing cases in classrooms. *Journal of Mathematics Teacher Education, 5*(4), 317-349.

Llinares, S., & Krainer, K. (2006). Mathematics (student) teachers and teacher educators as learners. In A. Gutiérrez & P. Boero (Eds.), *Handbook of research on the psychology of mathematics education: Past, present, and future* (pp. 429-460). Rotterdam: Sense Publishers.

Macnamara, O., Jaworski, B., Rowland, T., Hodgen, J., & Prestage, S. (2002). *Developing mathematics teaching and teachers: A research monograph.* Retrieved 8 February 2009 from
http://www.maths-ed.org.uk/mathsteachdev/pdf/mathsdev.pdf.

McDuffie, A. R. (2004). Mathematics teaching as a deliberate practice: An investigation of elementary pre-service teachers' reflective thinking during student teaching. *Journal of Mathematics Teacher Education, 7*(1), 33-61.

Murray, H., Olivier, A., & Human, P. (1999). Teachers' mathematical experiences as links to children's needs. In B. Jaworski, T. Wood, & S. Dawson (Eds.), *Mathematics teacher education: Critical international perspectives* (pp. 33-41). London: Falmer Press.

Phillips, E. (1997). Why might a teacher want to do research, and what is teacher research anyway? In V. Zack, J. Mousley, & C. Breen (Eds.), *Developing practice: Teachers' inquiry and educational change.* Geelong, Australia: Centre for Studies in Mathematics, Science, and Environmental Education, Deakin University.

Ponte, J. P. (1994). Mathematics teachers' professional knowledge. In J. P. Ponte & J. F. Matos (Eds.), *Proceedings of the eighteenth PME conference,* vol. 1 (pp. 195-210). Lisbon: University of Lisbon.

Ponte, J. P., & Chapman, O. (2006). Mathematics teachers' knowledge and practices. In A. Gutiérrez & P. Boero (Eds.), *Handbook of research on the psychology of mathematics education: Past, present, and future* (pp. 461-494). Rotterdam: Sense Publishers.

Reiss, C., Schoenfeld, A., & Törner, G. (2007). *Professional development of mathematics teachers: Research and practice from an international perspective.* Report 52/2007. Oberwolfach, Germany: Mathematicshes Forschungsinstitut Oberwolfach.

Remillard, J. T., & Geist, P. K. (2002). Supporting teachers' professional learning by navigating openings in the curriculum. *Journal of Mathematics Teacher Education, 5*(1), 7-34.

Romberg, T. A., & Carpenter, T. P. (1986). Research on teaching and learning mathematics: Two disciplines of scientific enquiry. In M. C. Wittrock (Ed.), *Handbook of research on teaching,* 3rd ed. (pp. 850-873). New York: Macmillan.

Rösken, B., & Törner, G. (2008). Mathematics done differently: An innovative approach to furthering professional development of German teachers. Paper presented at the Interactive Symposium, Teacher Academic Partnerships: International Approaches to Teacher Professional Development, Division

K, Section 8, Conference of the American Educational Research Association (AERA), New York.

Rowland, T., Huckstep, P., & Thwaites, A. (2005). Elementary mathematics teachers' subject knowledge: The knowledge quartet and the case of Naomi. *Journal of Mathematics Teacher Education, 8*(3), 255-281.

Schifter, D. (1998). Learning mathematics for teaching: From a teachers' seminar to the classroom. *Journal of Mathematics Teacher Education, 1*(1), 55-87.

Shavelson, R. J., Phillips, D. C., Towne, L., & Feuer, M. J. (2003). On the science of education design studies. *Educational Researcher*, 32(1), 25-28.

Shimizu, Y. (2008). The role of the outside expert in lesson study in Japan. Paper presented at the Interactive Symposium, Teacher Academic Partnerships: International Approaches to Teacher Professional Development, Division K, Section 8, Conference of the American Educational Research Association (AERA), New York.

Shulman, L. S. (1987). Knowledge and teaching: Foundations of the new reform. *Harvard Educational Review, 57*(1), 1-22.

Simon, M. (2008). The challenge of mathematics teacher education in an era of mathematics education reform. In B. Jaworski & T. Wood (Eds.), *The mathematics teacher educator as a developing professional,* vol. 4 of *International handbook of mathematics teacher education* (pp. 17-30). Rotterdam: Sense Publishers.

Skemp, R. (1976). Relational understanding and instrumental understanding. *Mathematics Teaching, 77,* 20-26.

Smith, A. (2004). *Making mathematics count: The report of Professor Adrian Smith's inquiry into post-14 mathematics education.* London: HMSO.

Steffe, L. P., & Thompson, P. W. (2000). Interaction or intersubjectivity? A reply to Lerman. *Journal for Research in Mathematics Education, 31*(2), 191-209.

Straesser, R., Brandell, G., Grevholm, B., & Helenius, O. (2004). *Educating for the future: Proceedings of an international symposium on mathematics teacher education.* Sweden: Royal Swedish Academy of Sciences.

Sullivan, P., & Wood, T. (2008). *Knowledge and beliefs in mathematics teaching and teaching development.* Vol. 1 of *International handbook of mathematics teacher education.* Rotterdam: Sense Publishers.

Sztajn, P. (2008). Caring relations in the education of practising mathematics teachers. In B. Jaworski & T. Wood (Eds.), *The mathematics teacher educator as a developing professional,* vol. 4 of *International handbook of mathematics teacher education* (pp. 299-314). Rotterdam: Sense Publishers.

Tirosh, D., & Wood, T. (2008). *Tools and processes in mathematics teacher education,* vol. 2 of *International handbook of mathematics teacher education.* Rotterdam: Sense Publishers.

Tzur, R. (2008). Profound awareness of the learning paradox (PALP): A journey towards epistemologically regulated pedagogy in mathematics teaching and teacher education. In B. Jaworski & T. Wood (Eds.), *The mathematics teacher educator as a developing professional,* vol. 4 of *International handbook of mathematics teacher education* (pp. 137-156). Rotterdam: Sense Publishers.

Wagner, J. (1997). The unavoidable intervention of educational research: A framework for reconsidering research-practitioner cooperation. *Educational Researcher, 26*(7), 13-22.

Wells, G. (1999). *Dialogic inquiry: Towards a sociocultural practice and theory of education.* Cambridge, UK: Cambridge University Press.

Wilson, S. M., & Berne, J. (1999). Teacher learning and the acquisition of professional knowledge: An examination of research on contemporary professional development. *Review of Research in Education, 24,* 173-209.

Wood, T., Jaworski, B., Krainer, K., Sullivan, P. & Tirosh, D. (Eds.) (2008). *International handbook of mathematics teacher education.* Rotterdam: Sense Publishers.

Zack, V., Mousley, J., & Breen, C. (Eds.) (1997). *Developing practice: Teachers' inquiry and educational change.* Geelong, Australia: Centre for Studies in Mathematics, Science, and Environmental Education, Deakin University.

Zaslavsky, O. (2008). Meeting the challenges of mathematics teacher education through design and use of tasks that facilitate teacher learning. In B. Jaworski & T. Wood (Eds.), *The mathematics teacher educator as a developing professional,* vol. 4 of *International handbook of mathematics teacher education* (pp. 93-114). Rotterdam: Sense Publishers.

Part 2

Professional Development in Practice of Mathematics Teachers

Chapter 2

The Professional Development of a Mathematics Teacher in the Role of a Subject Leader

Cláudia Canha Nunes

Introduction

Teachers nowadays work in a world marked by economical, political, and organizational changes. The nature and context of curriculum, teachers' roles, and in-service teachers' training needs are also confronting changes. This educational change requires simultaneous and articulated action among teachers' professional development, curriculum development practices, professional school culture, organization, and leadership processes (Hargreaves, 1998).

This chapter focuses on the professional development of one mathematics teacher, surprised by her recent election as "subject leader," as she engaged in this new role along the span of a year. Specifically, the chapter focuses on how this teacher dealt with this new challenge and the difficulties she faced while carrying out her new task, in the context of a school project developed by the mathematics teachers' group, which decided to undertake collaboratively their curriculum design and practice.

Collaboration and Leadership in a School Context

Collaboration is a favourable context in which to reinforce trust, which is necessary for innovation, as well as to improve effectiveness, reduce overload, situate certainty, and offer opportunities to learn (Boavida & Ponte,

2002; Hargreaves, 1998). Furthermore, reflection and collaboration promote learning (Hargreaves, 1998; Schön, 1993). Sharing experiences, undertaking analyses, and reflecting on practices can promote teachers' professional development and thus the development of a new professional school culture (Nunes & Ponte, 2008).

During collaborative work, teachers are not alone in dealing with an issue; they work with other teachers who face the same problem. This collaboration creates an empathy that provides energy to overcome frustration, isolation, and the feeling of not being able to deal with the issue. In collaboration, it is possible to build a strategic action in which everyone has a role or responsibility and works toward the same goals. The structure of collaborative work is demanding since all actors are necessary. The work does not demand complex logistics but requires real experience, devotion, and sharing concerns, knowledge, and expectations. This sharing work enables teachers to come up with better strategies to overcome difficulties and problems and creates a feeling of trust that transforms a problem into an added value (Boavida & Ponte, 2002; Hargreaves, 1998).

According to Boutinet (1990, p. 195), the methodology of projects came as "a possible answer to the challenges put on the educational system." If we assume that all projects are built on personal interests and expectations, on former experiences of the team members, then any project is unique because it comes from the participants' concerns about a particular situation developed in given space and time, within a social context. We also have to consider the complexity of the situations and the uncertainty that motivates the elaboration of a project. During the project work—within a context of professional problem solving and training—systematic reflection and collaborative process characterize the ongoing work.

One essential element of collaboration is leadership. Leadership requires listening to colleagues describe experiments and preserving professional relations. These elements can occur in a more formal setting (e.g., group working sessions) or in a less formal setting, as when a colleague asks the subject leader for an opinion or advice or wants to share a personal experience, a task, or a classroom episode (Field, Holden, & Lawlor, 2003). Each individual contribution of the group members makes the collaboration stronger. According to Stewart (1997), the leadership process necessarily has to be shared and symbiotic. Effective management of interpersonal relationships requires that the subject leader know how to maintain a leadership role within the group—how to make decisions, to express himself or herself, to listen, to question and give information, to stimulate discussion, to regulate interaction, share experience and knowledge, encourage and facilitate communication, create a climate of trust, and solve possible conflict (Boavida & Ponte, 2002; Field, Holden, & Lawlor, 2003; Stewart, 1997).

As Sergiovanni (2004) explains, the leader is an orchestra conductor and a manager of meaning. He or she identifies what is important and gives a sense of direction or purpose through the articulation of a compelling worldview. The *vision* (Bryman, 1996) and leader expertise and skill, in association with personal experiences and creativity of colleagues (Field, Holden, & Lawlor, 2003), are key elements of a successful project of professional development of teachers, thus creating a new professional school culture (Sergiovanni, 2004).

From this discussion, we can say that there is a strong connection between collaboration and leadership; in fact, there is no collaborative work without leadership, and there is no leadership without collaborative work (Hargreaves, 1998; Krainer & Peter-Koop, 2003; Sergiovanni, 2004).

Methodological Orientation of the Study

The methodology of this study is qualitative and interpretative (Bodgan & Bicklen, 1994; Erickson, 1986; Merriam, 1988), using a case study design (Ponte, 2006; Stake, 1994; Yin, 1989). One teacher, whom I will call Ana, belonging to a group of mathematics teachers involved in innovative collaborative work in a middle school, was followed over the course of one year. This choice was related to the fact that she had recently been elected as subject leader. Data collection occurred during the school year 2007-2008 and included field observations (Jorgensen, 1989; Ludke & André, 1986), with reports coming from working sessions and other field notes made in the researcher's diary (Adler & Adler, 1994; Smyth, 1989), and an interview with the teacher (Fontana & Fey, 1994; Patton, 2002; Stake, 1995).

The observations include different aspects of the teacher's activity, especially attending meetings and undertaking classroom work. The role of the researcher in the study is an external one, of participant observer, in the working group sessions and in some classroom work (Adler & Adler, 1994).

According to the research plan, a preliminary data analysis takes place simultaneously with data collection. In this way, data analysis can inform the need for gathering new data. The data analysis is conducted at three levels. The first level has to do with organization of the collected data, such as the transcripts of group meetings, documents, and fieldwork notes. The second level involves the development of categories and classes that provide an interpretation of the data. The third level tries to explain the meaning of the data to provide input into the understanding of the phenomenon under study (Merriam, 1988).

The Mathematics Subject Group

The team was composed of fourteen grades 7 to 12 mathematics teachers. The group carried out several school projects in mathematics during the past decade. Most of these projects were inspired by teachers' need to improve their own practices or to overcome students' difficulties, trying to improve students' learning. During the school year 2007-2008, the project entitled Investigation, Proof, and Problem-Solving Tasks in Curriculum Management and Textbooks was carried out from grades 7 to 12.

During the working sessions, the mathematics group discussed tasks, constructed criteria to assess students' work, and reflected on the results of students. Also, there was space to share classroom experiences and discuss grading of students' work on their achievements and how to give students feedback, reflecting on the contributions of all colleagues and aspects of the practice.

This group is a very dynamic one, and its members usually participate in professional meetings. This dynamic is based on a particular team's concern to always update professional knowledge and be aware of educational changes. There are friendly interpersonal relationships among team members, and the working group environment is healthy and inviting. However, each group member has a different assigned status and role. For example, when following the group, it was possible to identify that one member, Simon, was recognized by the others as the catalyst and identified as a charismatic leader by the school community.

Some Results

Engagement as a Subject Leader by Ana

Ana is thirty-nine years old. She has been a teacher for twelve years, and she has a master's degree in mathematics. Her capacities and work are recognized by all the teachers of the group and by the school community. Some elements, such as the results achieved by her students in the national exam and the number of her students who enter university, contribute to this social recognition.

At the beginning of the school year 2007-2008, Ana was elected, by her colleagues, as the subject leader of the mathematics department. Her colleagues mention that her "first reaction was panic because she [felt] that she was not prepared to face the challenges and she did not know all the tasks that [come] with this job" (researcher's diary).

During the first two weeks of her work, Ana became aware of her duties. As a subject leader, she has to plan, prepare, and coordinate the group meetings, the project's working sessions, and the activities in which the group participates in the school; she also has to supervise classroom practices of her colleagues, evaluating them and writing assessment reports that will influence who advances in their careers.

Ana shows a great sense of responsibility toward the tasks that come with the job. This is why she is very anxious and worries about her performance, especially her assessment role as subject leader.

> Coordinating group activities wouldn't be very difficult for me. But I confess that I am not prepared to take such a task; there are a lot of things that I don't know. The worst of all is that I have to evaluate my colleagues that I regard as competent teachers and about whom I can only speak well. Also, I have to choose the best one of them. That makes me feel very anxious and nervous.
>
> Someone told me that my task is to evaluate them and write an assessment report which will influence the decision about who will progress in the professional career. (Interview)

Another reason for her uncomfortable feeling is that Ana was taken by surprise in the outcome of voting in the election of the subject leader. She explained in an interview, "I confess that I thought that there were other colleagues, Simon or Sebastian, [who] could be the elected one for this job ... because they have a larger experience than me, as they have been subject leaders, and they made a good job. They both are good coordinators."

Ana is a strong and determined person. When she has problems or doubts, she is not afraid to ask for help. She knows that it is important to have this experience of coordinating the mathematics subject group, even though she thinks that she is not a leader. Another issue is the fact that Ana is one of the youngest teachers in the mathematics group, a group in which her mathematics teacher took part when she was a high school student in that school.

> There are some people that have better leader skills. This task is mostly about managing interpersonal relationships, and I don't like to do that. I like students and dealing with them. For instance, I have older colleagues in the mathematics' subject group, and I have to tell them, "You have to do this ..., you have to do that [sigh] ..." One of my colleagues has been my teacher! It feels awkward. (Interview)

Ana has good relationships with all of her subject group colleagues. However, she recognizes that she deals differently with the two older colleagues of the group. An example of this issue is when Ana had to explain to them their new tasks, particularly the role of teacher partner in "study lesson" classes:

> "The two older colleagues were not present in the group meeting where we planned and decided these situations, and they aren't used to hav[ing] another colleague with them in the classroom, helping them. They are finishing their professional careers, and they never had such experiences, that's the reason why I was more careful explaining [to] them how and why" (interview).

As a teacher, Ana mentioned, she likes to follow her colleagues' practices to be well informed and to learn from their experiences; as a subject leader, she believes that she has to know what her colleagues are doing as teachers "so that I can defend them in school pedagogical plenaries and with principals" (interview).

Project Working Sessions

As a mathematics teacher, Ana needs to share work with her colleagues. As a subject leader, she considers this to be a "special" group in which all the work is planned by the team and in which there is a strong reflexive attitude, all members sharing with each other their own practices and experiences. "This is a special group. We work together for a long time. I hope it will stay like this! This is a special group, because I have already worked at four different schools, and this one is the first where the mathematics' subject group works in collaboration, sharing all the tasks" (interview).

Since the beginning of the school year, Ana showed some small changes in how she participated in and contributed to the project's working sessions. She intervened more, enriching group discussions and reflections with her suggestions and ideas. Ana revealed that she has a capacity for sharing, listening, and integrating colleagues' suggestions, especially those that come from Sebastian and Simon. The subject for the project group at e-working platform *moodle* highlights that fact, as all group members are teachers at *moodle*. Thus, all teachers in the group have the opportunity to create discussion forums or to publish didactic materials (in particular, tasks and tests), which can be enriched and improved without having to wait for the working group sessions.

Ana refers to Simon as the catalyst of the working processes developed in the mathematics subject group and in the different projects developed by the group. "He has his own beliefs about how professional and school culture should be—with strong collaborative work and a continuous development and learning attitude—and the entire group follows his *vision*" (researcher's diary). Furthermore, and besides the fact of being the subject leader, Ana decided to share coordination of the school project. She looks at collaboration as a natural working situation of the subject group and as a tool for help, support, and sharing: "I am always telling what happens in my classes, and I like to know my colleagues' experiences so I can have different opinions on my decisions and classroom strategies" (interview).

Project sessions have three different moments of work: a first moment of plenary work; a second moment of subgroup work, where teachers' work is done by school grade; and a third moment of making decisions and attributing individual tasks until the next project working session. Usually, it is Ana who coordinates the first and third moments. However, during all working sessions, everyone is invited to share the leadership; to give opinions and make suggestions; and to participate with ideas, tasks, materials, classroom strategies, and experiences, in a trusting and learning environment. As Ana said in an interview, reporting her subgroups' work, "'Look, I did this, and I think it is interesting. Look, this [task] didn't go well. I did this task, and my students didn't get the answer; we really have to reflect on and change it.' So I like to share these things with my colleagues and my own experiences."

In project working sessions, the group works collaboratively in curriculum management. In subgroups of two or three teachers, they start planning using the textbook and following the program unit. They look carefully at the textbook's tasks, solve them, and then decide which are the more interesting and rich for students to solve. If necessary, they construct a worksheet with investigations, proofs, or problem-solving tasks to help students improve their learning.

Ana recognizes that projects are a fundamental strategy for the professional development of teachers and for the development of professional and school culture: "With these projects, people always learn something, [and] they are always enriching in some way" (interview). On the other hand, if there were no projects, "the group would continue being a group, but a much poorer one, I think. Also, our task would be much more impoverished" (interview).

This culture of collaboration seems to have been fostered by the ways in which almost all teachers of the group have participated with enthusiasm. They appear to think that these initiatives are essential to their growth as teachers. These initiatives seem to be the key to how they work as a group and have contributed to their working culture, where exchanges of ideas, experiences, and materials are welcome.

Some Reflections

Ana thinks that she still has much to learn. But the context in which she develops her profession, in particular the subject group that she leads, represents an advantage for her. As one of the younger members of the group, when she was elected subject leader, she chose to have a humble and learning attitude. Every time Ana needs help, she asks her colleagues, and she can count on them for collaboration and experience, mostly from Simon. The interpersonal relationships and working group environment seem to be key elements for Ana's professional development and leadership skills, not only as a teacher but also as a subject leader (Hargreaves, 1998; Krainer & Peter-Koop, 2003; Sergiovanni, 2004).

The mathematics subject group is a cohesive and collaborative group, teachers in it sharing and reflecting on their own practices. These characteristics represent two catalysts that facilitate and reinforce trust and a collective feeling of support, which are necessary for educational innovation and change (Boavida & Ponte, 2002; Field, Holden, & Lawlor, 2003; Hargreaves, 1998; Stewart, 1997).

Finally, the various initiatives of the group, in particular its projects, are a key to the sustainability of the culture of collaborative work (Nunes & Ponte, 2008). This dynamic and working context seem to motivate the involvement of teachers in teaching and learning. In particular, such a dynamic appears to support the professional development of Ana and her capacity to accept new challenges. The professional practice of these teachers, supported by this working environment, shows that current curriculum orientations can be implemented not just at an individual or small group level but also by a whole school mathematics subject group.

From this study, new issues emerge for future research. How do teachers' practices and curriculum management influence students' learning of mathematics? What conditions are necessary at schools, and more widely in the social context, so that this kind of collective curriculum management takes place in line with current curriculum orientations? What is the role of leadership in this process?

References

Adler, P., & Adler, P. (1994). Observational techniques. In N. Denzin & Y. Lincoln (Eds.), *Handbook of qualitative research* (pp. 377-391). Thousand Oaks, CA: Sage.

Boavida, A. M., & Ponte, J. P. (2002). Investigação colaborativa: Potencialidades e problemas. In GTI (Ed.), *Reflectir e investigar sobre a prática profissional* (pp. 43-55). Lisboa: APM.

Bodgan, R., & Bicklen, S. (1994). *Investigação qualitativa em educação: Uma introdução à teoria e aos métodos.* Porto: Porto Editora.

Boutinet, J. P. (1990). *Antropologia do projecto.* Lisboa: Instituto PIAGET.

Bryman, A. (1996). Leadership in organizations. In S. Clegg, C. Hardy, & W. Nord (Eds.), *Handbook of organization studies* (pp. 276-292). London: Sage.

Erickson, F. (1986). Qualitative methods in research on teaching. In M. C. Wittrock (Ed.), *Handbook of Research on Teaching* (pp. 119-161). New York: Macmillan.

Field, K., Holden, P., & Lawlor, H. (2003). *Effective subject leadership.* London: Routledge & Falmer.

Fontana, A., & Fey, J. (1994). Interviewing: The art of science. In N. Denzin & Y. Lincoln (Eds.), *Handbook of qualitative research* (pp. 377-391). Thousand Oaks, CA: Sage.

Hargreaves, A. (1998). *Os professores em tempos de mudança: O trabalho e a cultura dos professores na idade pós-moderna.* Lisboa: McGraw-Hill.

Jorgensen, D. (1989). *Participant observation.* Newbury Park, CA: Sage.

Krainer, K., & Peter-Koop, A. (2003). The role of the principal in mathematics teacher development. In A. Peter-Koop, V. Santos-Wagner, C. Breen, & A. Begg (Eds.), *Collaboration in teacher education* (pp. 169-190). Dordrecht: Kluwer.

Ludke, M., & André, M. (1986). *Pesquisa em educação: Abordagens qualitativas.* São Paulo: EPU.

Merriam, S. B. (1988). *Case study research in education: A qualitative approach.* San Francisco: Jossey-Bass.

Nunes, C. C., & Ponte, J. (2008). Os projectos de escola e a sua liderança. In GTI (Ed.), *O professor de matemática e os projectos de escola* (pp. 11-37). Lisboa: APM.

Patton, M. (2002). *Qualitative research and evaluation methods*. London: Sage.

Ponte, J. P. (2006). Estudos de caso em educação matemática. *BOLEMA, 25,* 105-132.

Sergiovanni, T. J. (2004). *O mundo da liderança: Desenvolver culturas, práticas, e responsabilidade pessoal nas escolas*. Porto: ASA.

Shön, D. A. (1993). *The reflective practitioner: How professionals think in action*. Aldershot: Avebury.

Smyth, J. (1989). Developing and sustaining critical reflection in teacher education. *Journal of Teacher Education, 30*(2), 2-9.

Stake, R. E. (1994). Case studies. In N. Denzin & Y. Lincoln (Eds.), *Handbook of qualitative research* (pp. 236-247). Thousand Oaks, CA: Sage.

Stake, R. E. (1995). *The art of case study research*. Thousand Oaks, CA: Sage.

Stewart, H. (1997). Metaphors of interrelatedness: Principles of collaboration. In H. Christiansen, L. Goulet, C. Krentz, & M. Maeers (Eds.), *Recreating relationships: Collaboration and educational reform* (pp. 27-53). Albany: State University of New York Press.

Yin, R. K. (1989). *Case study research: Design and methods*. Newbury Park, CA: Sage.

Chapter 3

Mentoring as Professional Development: A Case from Secondary-Level Mathematics

Ginger Rhodes and Patricia S. Wilson

Introduction

Learning a profession requires formal study and practical experience, and teaching is no exception. In the United States, field experiences have become an integral part of teacher preparation programs, and university students eagerly embrace the opportunity to work in schools and learn from veteran teachers. Practising teachers might be involved in formal study at a university or school district program, but we often overlook their need for practical experiences, believing that their daily teaching fills this role. In-service teachers can learn from their daily practice, but we argue that professional development that intentionally uses their practice to stimulate reflection provides more opportunities for substantial learning. Teachers need structured opportunities for reflecting on their practice, focusing on student learning, and discussing their insights with colleagues. We have found that professional development set in the context of mentoring a student teacher or novice teacher can help practising teachers think about their own teaching in new ways.

In the United States, teachers are rarely provided with time during the day to intentionally reflect on their practice, but it is a common practice for experienced teachers to mentor a student teacher or novice teacher. Traditionally, the mentor is expected to dedicate a significant amount of time and effort, and the novice teacher is considered to be the learner. We want to challenge that notion and argue that the opportunity for mentor learning can increase by shifting the responsibility of the mentor from providing advice to undertaking *collaborative work with a protégé in order to learn.*

Mentoring a student teacher or colleague is difficult to do well (Sudinza, Giebelhaus, & Coolican 1997), and often the responsibilities are not well defined (Carver & Katz, 2004). Some mentor teachers seek to model good teaching, and others prefer to guide the practice of their protégés by posing challenging questions (Wilson, Anderson, Leatham, Lovin, & Sanchez, 1999), but few think of their mentoring as an opportunity for their own learning. Although teachers claim to learn how to teach from experimenting with their practice and talking with colleagues (Wilson, Cooney, & Stinson, 2005), the structure of most mentoring programs is not built on collaboration. By definition, the mentor has experiences that the protégé lacks. Traditional mentoring experiences set up power relationships among mentors, university supervisors, and student teachers that discourage collaboration and learning from each other. We found that, by altering the structure of traditional field experiences for student teachers and the responsibilities of the participants, we could create professional growth experiences for all participants—student teachers, mentor teachers, and university teachers.

In the following sections, we discuss our efforts to use mentoring experiences as professional development specifically for in-service teachers but note that the professional development was for all participants. We built our program on the theory that learning *from* practice while being engaged *in* practice is extremely beneficial *for* practice (Stein, Smith, & Silver, 1999). We incorporated this theory as we designed experiences for participants to learn mathematics and strategies for teaching mathematics. We based this chapter on our experiences with mentors of students who were preparing to be secondary mathematics teachers. Specifically, we describe opportunities for mentors to learn both mathematics and pedagogy from their mentoring practice, and we identify three particular characteristics of our program that encouraged and supported professional growth.

Professional Growth as Part of Mentoring

Partnerships in Reform in Mathematics Education (PRIME) is a multilevel professional development project for three groups of teachers: student teachers preparing to be high school mathematics teachers, practising teachers who serve as mentors, and university teachers who teach courses for the student teachers. The goal of PRIME is to improve mathematics teaching at the secondary level by providing opportunities for pre-service teachers, in-service teachers, and university teachers, as a collective, to examine the practice of teaching mathematics and to discuss the mathematical knowledge and skills needed for teaching. PRIME has been part of the mathematics teacher education program at the University of Georgia since 1998, and it was founded on the idea that the context of student teaching could provide meaningful learning opportunities not only for student teachers but also for university teachers and mentor teachers. Student teachers were placed in groups of two to five at each participating school. In most cases, one student teacher was paired with one mentor teacher, but in some cases there were two mentor teachers for one student teacher or two student teachers with one mentor. The community of learners, which we denoted as a *cluster,* included the mentor teachers, student teachers, and university teacher at each school site. The cluster met as a collective for thirty to forty-five minutes once a week to examine and discuss *the work of mathematics teaching.* We used this phrase to emphasize a focus on mathematics. All members of the community were encouraged to participate in the cluster meetings. The content of these meetings depended on events in the mathematics classes and the leadership of the university teacher who served as a facilitator. At some meetings, the university teacher would initiate discussions by sharing interesting mathematical and pedagogical questions that occurred in various classrooms, and in other meetings the university teacher would encourage mentor teachers and student teachers to initiate the discussions. Most importantly, the focus of the cluster meetings centred on the mathematical activities that took place in the teachers' classrooms.

The weekly cluster meeting offered a rare opportunity for mentor teachers to discuss their practice with a group of knowledgeable colleagues. We encouraged members of the cluster to move beyond the traditional mentor/student teacher dyad and begin to work as a collaborative group to enhance their teaching of mathematics. University teachers helped to focus the group on issues related specifically to teaching mathematics rather than general issues of teaching. The cluster discussions were valuable for all participants when the discussion centred on events from the teachers' mathematics classrooms and specifically on the mathematics in the lesson. Some examples of activities included examining student work, discussing a student's mathematical question, and solving a mathematics task. The separate school clusters came together four times during the year to share ideas. These meetings included mathematical activities (e.g., analyzing student work or discussing a videotaped mathematics lesson) and opportunities for the teachers to discuss their teaching and mentoring practices with colleagues from other schools.

Characteristics of Learning Opportunities

After studying PRIME at work in thirteen schools, we offer an example of a situation in which we observed great mentor participation and enthusiasm and then discuss three characteristics that promoted reflection, by mentors, on their practice. During a cluster meeting, a group of three mentor teachers, three student teachers, and the university teacher watched a ten-minute videotaped lesson on synthetic division from one of the student teacher's classes. The university teacher encouraged the student teacher to share a video episode from her class, and the student teacher took the leadership role by selecting a class episode and bravely sharing her teaching. Before sharing the video, the student teacher described her concern that she had not let the students become involved in thinking about the mathematics; instead, she had led them through a step-by-step process of synthetic division without allowing them to develop it or question it. Her stated concern focused the group's observation, so everyone watched the video with a purpose.

After watching the class episode, the teachers immediately began discussing what they had observed. The experienced teachers were able to recognize subtle student questions that had not seemed significant to the student teacher. The group was attracted to one question indicating that the student was looking at a relationship between two numbers in the algorithm. One mentor called his comment insightful, claiming that it was very difficult for students to make such connections. As a group, the teachers focused on this student's mathematical thinking and recognized that his observation had provided an opportunity to teach mathematical connections even when the lesson was about a procedural task. Another student teacher asked, "Why do we teach synthetic division?" Various responses to this question led to a discussion of the use of synthetic division in factoring polynomials of degrees greater than two and then jumped to the relationship between synthetic division and long division. The group began to focus on connections among mathematical topics such as critical points of a function and derivatives. The conversation shifted from the mathematics of high school students to the mathematics of the teachers, and they began to see implications for the curriculum.

This example of a cluster meeting discussion illustrates the excitement that can be generated by using actual episodes from practice and focusing on the mathematical thinking of students. The discussion jumped around because of the spontaneous interest and participation of teachers who taught a variety of classes. A discussion that began with a procedural task developed into a collection of mathematical insights. Each cluster member participated in the analysis of the video, learning mathematics, seeing new ideas for teaching mathematics, and contributing to the conversation. This kind of collaboration fostered mathematical thinking and insights into students' mathematical thinking, which consequently helped each teacher to reflect on critical features of practice.

Characteristics of good professional development include working within the context of practice, focusing on students' mathematical thinking, and collaborating with colleagues (Garet, Porter, Desimone, Birman, & Yoon, 2001; Hiebert, Gallimore, & Stigler, 2002; Loucks-Horsley, Love, Stiles, Mundry, & Hewson, 2003; Stein, Smith, & Silver, 1999). These characteristics were intentionally designed into the PRIME format. We chose to discuss these three characteristics because they were prevalent in successful cluster meetings, and they suggest ways that professional development can be a product of mentoring.

Learning within the Context of Practice

Mentors and student teachers were enthusiastic about the discussion of synthetic division because it was related to their practice as mathematics teachers. By structuring events within the context of practice, we were assured that the work reflected the excitement, constraints, and reality of teaching mathematics. Learning from practice can take place within a mathematics classroom, but there are additional venues that tap into practice. Studying authentic student work, watching videotapes of a mathematics lesson, creating or selecting a mathematics problem, or designing a lesson and an assessment are all examples of activities that can allow participants to study mathematics and teaching within the context of practice. Learning within the context of practice involves identifying key issues, solving problems, connecting ideas, and reflecting on experiences. We know that teaching is complex, and learning within practice acknowledges this complexity. The goal of learning within a context is not to emulate a master teacher or replicate a successful lesson. The point is to study the teaching and its influence on student learning so that we understand why it was successful. Rather than a mentor modelling good practice for a student teacher, all cluster members reflected on the teaching episode and could apply ideas to their own teaching.

Learning from Students' Mathematical Thinking

The participants in the example discussed a student's mathematical thinking, which prompted them to share their own mathematical thinking. They had made strides in the difficult task of switching the focus of their study from "What should I do as a teacher?" to "What are my students thinking, and how can I build on their thinking?" If someone is working to improve his or her teaching of mathematics, a naive approach might be to try to identify good teaching skills and learn to execute those skills smoothly and even creatively. This mastery of skills is appropriate for some tasks, but it is not sufficient for the complex task of teaching. It is not possible for teachers to help students learn mathematics if they do not understand the mathematical thinking of their students. Many teachers value getting to know their students, their interests in and out of school, their family backgrounds, and their aspirations (Wilson, Cooney, & Stinson, 2005). Although this kind of knowledge can be a valuable asset for teaching, it is not synonymous with understanding students' *mathematical* thinking. Good teaching is dependent on the ability to assess, understand, and influence students' mathematical thinking. In other words, teachers need to know how to elicit students' mathematical thinking through clever assessment items, class discussions, and insightful questions. Teachers need a deep understanding of mathematics in order to interpret what students reveal about their mathematical thinking. Rather than a mentor and protégé focusing exclusively on what a teacher should do, a collegial investigation into what students are thinking facilitated reflection on mathematics, curriculum, and mathematical connections.

Learning from Collaboration

The cluster meeting began with a concern from one student teacher, which quickly became the concern and focus of the entire group. We have found that collaboration is a key characteristic that distinguishes a mentor who is learning from the mentor who is advising. The collaborative process is about creating something new, and a group that truly collaborates can attest to the non-mathematical statement that the whole is greater than the sum of its parts! Teachers frequently work together to produce a product, such as curriculum for a common class or a departmental exam. Although this work is important and might lead to teacher learning, the cooperative work that allows each teacher to contribute does not necessarily require a teacher to rethink ideas or change. We argue that productive collaboration includes joint work that causes teachers to reconsider their thinking and practice. As mentors began to discuss students' mathematical thinking and address complex issues that came from their daily practice of teaching mathematics, they began to value and learn from the collaboration. In our example of a cluster meeting, the mentors and protégés contributed by their expertise to analyze a situation that had been posed to the group, and they thought about how the shared ideas could play out in their own practice. In collaborative work, each person not only contributes but also walks away with valuable insights and ways to act on those insights. Collaboration is difficult to achieve because it is not just a characteristic of a program but also a disposition of the participants. Through collaboration, participants jointly construct new ideas. Professional development that encourages collaboration within the group can promote mentor reflection and learning.

Improving Opportunities for Mentor Learning

PRIME encouraged student/mentor dyads to collaborate and examine the work of teaching with a focus on students' mathematical thinking. The dyads planned lessons together, taught lessons together, and examined quizzes and tests together. Although we believe that all of these activities were valuable for both the mentors and the student teachers, we claim that professional growth of the mentors depended on the extent to which these activities incorporated the three characteristics—learning from the context of practice, learning from students' mathematical thinking, and learning from collaboration.

To highlight our claim, we share a scenario that is frequently observed in student teaching and contrast it with a scenario that incorporates the three characteristics to a greater extent. We argue that the second scenario provides more opportunities for mentor reflection and ultimately more opportunities for teacher learning.

We often saw mentor teachers sharing knowledge about student difficulties, mathematical connections, and teaching strategies with their student teachers. Although the mentor teacher's advice can be helpful to the student teacher, the opportunity for mentor learning is limited by the kind of interaction between the two teachers. In sharp contrast, we observed a mentor teacher who wanted to use a mathematics task in class that she had never used before. The mentor teacher and student teacher conceived the task together before they created their lesson plan. While completing the task, both teachers commented on the emerging mathematical ideas and potential mathematical connections. They also discussed the aspects of the activity that were difficult for them, and they considered aspects that might be difficult for students. The teachers made reference to what students already knew about the underlying mathematics of the task. After the teachers completed the activity, they went back to consider how to introduce the activity.

On the surface, both examples include the three characteristics that promote learning, but we claim that the second example is considerably different from the first. The teachers collaboratively considered how their actions as teachers might affect student learning during the task. The teachers completed the activity themselves and made connections between students starting points, potential student solution paths, and possible student struggles. The mentor and student teacher both took an active role in planning a lesson; both teachers shared their ideas, listened to and critiqued each other's ideas, and created the lesson jointly. Furthermore, both teachers approached the activity with an investigative stance.

Facilitating Mentor Reflection

Mentoring can provide a unique opportunity for teachers to collaborate and discuss the work of mathematics teaching. One PRIME mentor teacher made the following comment about her experience:

> Sometimes I just need a jumpstart to be more reflective about my teaching and to take more risks. It's very easy to get into a comfortable mode of doing the same things each year. This year has challenged me to rethink what I am teaching, why I am teaching it, and if there are better ways to get my students to learn the things I want them to learn.

As the mathematics education community explores new ways to provide professional development for in-service teachers, we need to consider learning opportunities that exist within current structures. PRIME is an example in which hosting a pre-service teacher provided a context for mentor teachers to investigate their own teaching practices. We have offered examples of how mentoring activities can provide valuable opportunities for reflection for both the novice and the mentor teacher. In our concluding remarks, we want to emphasize that teachers need support to carry out this mentoring work. Co-planning, co-teaching, and jointly analyzing and creating mathematical tasks are not common activities among teachers. The three characteristics that we have highlighted in our example present challenges for both the mentor and the student teacher.

First and foremost, both the mentor and the student teacher must approach their joint work as learners. They must be willing to investigate their practice, share their own ideas, and consider each other's ideas. We found that, when a teacher dominated the discussion by sharing previous experiences without including the student teacher in the analysis of the situation, the learning experience for the student teacher was reduced to listening, and the learning experience for the teacher was limited to self-reflection. Often this appeared in cluster meetings when the mentor teachers offered advice to the student teacher rather than reflecting on practice and considering additional viewpoints. Teachers must be willing to reflect on their practice and value learning from their colleagues. This is not an easy task for teachers to accomplish.

Second, limited time is a constraint that many teachers deal with daily. For the mentor teacher to grow professionally from her mentoring work, she must work with the student teacher, and that effort requires significant time. Together the mentor and student teacher must carve out time to dedicate to their joint work.

These challenges need careful attention when organizing and planning mentoring activities. We found that the university teacher's role was instrumental in this work. For example, facilitating meetings, encouraging inquiry, participating in discussions, and designing student teacher assignments were vital contributions in the process of transforming field experiences into professional development opportunities. The power relationships and differentiated areas of expertise in traditional field experiences interfere with collaboration. If the mentor teacher is considered the expert practitioner, the university teacher is considered the supervisor and organizer of the field experience, and the student teacher is considered the learner, then the field experience context is not conducive to sharing, listening, and learning. It is not the purview of the university teacher to "give voice" to the mentor teacher or the student teacher; rather, the responsibility of the facilitator is to foster collaboration by ensuring that each voice can be heard, acknowledged, and thoughtfully considered. As teacher educators, we need to consider our role within mentoring structures. We need to explore productive strategies that work within field experiences to support mentor teachers as they learn mathematics and strategies for teaching mathematics. We believe that utilizing current educational structures, such as field experiences, in strategic ways will help teachers to extend their mathematical and pedagogical knowledge. Ultimately, we need to design experiences that promote learning within practice, that emphasize learning from students' mathematical thinking, and that facilitate true collaboration.

References

Carver, C. L., & Katz, D. S. (2004). Teaching at the boundary of acceptable practice: What is a new teacher mentor to do? *Journal of Teacher Education, 55,* 449–462.

Garet, M. S., Porter, A. C., Desimone, L., Birman, B. F., & Yoon, K. S. (2001). What makes professional development effective? Results from a national sample of teachers. *American Educational Research Journal, 38* (4), 915–945.

Hiebert, J., Gallimore, R., & Stigler, J. W. (2002). A knowledge base for the teaching profession: What would it look like and how could we get one? *Educational Researcher, 31*(5), 3–15.

Loucks-Horsley, S., Love, N., Stiles, K., Mundry, S., & Hewson, P. (2003). *Designing professional development for teachers of science and mathematics.* 2nd ed. Thousand Oaks, CA: Corwin Press.

Stein, M. K., Smith, M. S., & Silver, E. (1999). The development of professional developers: Learning to assist teachers in new settings in new ways. *Harvard Educational Review, 69* (3), 237–269.

Sudinza, M., Giebelhaus, C., & Coolican, M. (1997). Mentor or tormentor: The role of the cooperating teacher in student teacher success or failure. *Action in Teacher Education, 18,* 23–25.

Wilson, P. S., Anderson, D. L., Leatham, K. R., Lovin, L. H., & Sanchez, W. B. (1999). Giving voice to mentor teachers. In F. Hitt & M. Santos (Eds.), *Proceedings of the twenty-first annual meeting of the North American Chapter of the International Group for the Psychology of Mathematics Education,* vol. 2 (pp. 811–817). Cuernavaca, Mexico: Universidad Autonoma del Estado de Morelos.

Wilson, P. S., Cooney, T., & Stinson, D. (2005). What constitutes good mathematics teaching and how it develops: Nine high school teachers' perspectives. *Journal of Mathematics Teacher Education, 8,* 83–111.

Chapter 4

Facilitating Mathematics Teachers' Professional Development through Ranking and Promotion in Mainland China

Yeping Li, Rongjin Huang, Jiansheng Bao, and Yadong Fan

Introduction

The search for effective approaches to achieving excellence in mathematics education has led to an increased interest for teaching and teacher education practices in some high-achieving educational systems, including in China (Li, 2007; Li, Ma, & Pang, 2008). In particular, recent investigations revealed remarkable differences between American and Chinese teachers in their mathematics knowledge for teaching (Ma, 1999). Paradoxically, Chinese mathematics teachers often receive less formal education than their American counterparts (Ma, 1999). A possible explanation of the differences is that Chinese teachers continue to improve their knowledge and skills for teaching over the years (Li & Huang, 2008; Li, Ma, & Pang, 2008). Although it is recognized that China has a coherent teacher professional development system (Stewart, 2006), few studies have examined what can happen in this system. How can Chinese teachers experience different professional development compared with their counterparts in other educational systems, such as in the United States? In this chapter, we centre on the ranking and promotional system of teachers used in China to facilitate professional development and on the cultural context that supports this practice.

Teaching as a Profession in China: A Brief Background

With a long history of education, China has a deeply rooted cultural value placed on teachers and teaching. Chinese teachers are respected for their in-depth knowledge and exemplary moral traits, a conception that presents more differences than similarities compared with the conception of a good teacher in the West (e.g., in England; see Beishuizen, Hof, van Putten, Bouwmeester, & Asscher, 2001; Jin & Cortazzi, 1998). Teachers' work accomplishments are officially recognized according to rank and title associated with years of experience, capability of teaching, and level of instructional leadership. During the past decade, the country experienced rapid development in economy and technology, so that the Chinese educational system also strives to regulate and upgrade its teaching workforce for both pre-service preparation (Li, Zhao, Huang, & Ma, 2008) and in-service professional development (Huang & Li, 2008).

In 1994, release of the Teachers Act (Ministry of Education, 1994) was a crucial event that identified a teacher's position and responsibility from the perspective of the law. Later, release of the Regulations of Teacher's Qualification (Ministry of Education, 1995) further shaped the system of teachers' professional qualification. According to the regulations, the minimum education required for elementary school teachers is completion of a three-year program offered by normal schools for junior secondary school graduates; the minimum for junior high school teachers is completion of a three-year program offered by junior normal colleges for senior high school graduates; and the minimum for senior high school teachers a four-year bachelor degree. Over the past two decades, this three-stage normal education was established in Mainland China, and it has contributed to the preparation of teachers from elementary to secondary school levels.

According to the promotional system of secondary school teachers, the positions include senior-rank teacher, secondary teacher level 1 (intermediate), and secondary teacher level 3 (primary). For each level of teacher position, specifications in terms of political, moral, and academic qualifications are clarified. For example, a secondary school teacher at level 2 should meet political and moral qualifications; have completed a three-year program after senior high school; have completed one year of probation and an examination evaluating basic knowledge of education, psychology, pedagogy, and mathematical education; and be able to show that he or she can teach at least one course at junior high school. The requirements for being a teacher at the senior rank include (1) at least five years of experience as a secondary school teacher at level 1 or a PhD degree and (2) demonstration of the ability to take the responsibility of teaching as a "senior" secondary teacher. The second condition is further specified as follows: (a) the candidate should have systematic and sound academic training and specialized knowledge, should have plentiful teaching experience to be able to teach effectively, or should be specialized in political and moral education and classroom management, having achieved high performance and obtaining rich experience; (b) the candidate should engage in secondary school education research and teaching and be able to write teaching experience summaries, scientific reports, and research papers that integrate theory and practice at certain academic levels or have made remarkable contributions to the improvement of other teachers' academic levels and teaching abilities.

The Conception and Strategies of Teacher Professional Development in China

The Conception

The conception of teacher professional development in China shares some similarities but also some important differences with the conception of professional development in a Western context (Li, Kulm, & Smith, 2006). Although teachers' professional development is commonly perceived as improving the knowledge and skills needed for teaching, there are big cross-cultural differences related to the needs we perceive and the way we can integrate them as part of teachers' professional development. For example, as their professional development in the United States often takes the form of workshops and recommendations, it can be far from their instructional practice, not always appropriate, and even contradictory to their practice (Guskey, 2003). In contrast, teachers' professional development in China is conceived as activities that are practical. Rather than lengthy recommendations and workshops, professional development in China can happen in a teacher's daily life in a way that is pertinent to his or her needs.

This cross-cultural difference in professional development can be related to what is perceived as the professional nature of teaching. Teaching is conceived of in China as a professional practice, not private and personal. Common beliefs and values guide what teachers need to do for students (*"Jiao Shu Yu Ren,"* a special phrase meaning that a teacher is expected not only to teach students knowledge but also to educate them). A model of teaching behaviour is often established informally through frequent exchanges of ideas among teachers inside and outside their own schools (Huang & Bao, 2006). In particular, master teachers[1] in China are often identified as those who bear most of the culturally valued moral character and expertise to be followed by others. Teachers can learn from others the types of knowledge and skill that are publicly valued and locally proven to be effective. Teachers are conscious of the necessity of learning, and they know what they can learn from others through their daily teaching activities. Their expertise, reflected as knowledge and skills needed for and in their mathematics teaching, is real. There is recognition of a difference in expertise from one teacher to another.

The Strategies

Taking professional development to be independent from classroom-based instructional practice, American teachers receive information from professional publications about what can be considered good instructional practice and what is needed for achieving it. Many large-scale professional development programs and workshops are also available to support teachers' professional growth and change. Thus, American teachers' knowledge and skills for teaching are often enhanced through an approach such as "learning from being told and by doing" (Li, Kulm, & Smith, 2006). Although this approach has its own advantages, it is not clear that teachers perceive the need to engage in such professional development (Li & Kulm, 2008). In contrast, Chinese teachers' knowledge and skills for teaching are often improved through an approach such as "learning from example and by doing." Such strategies have been in place for many years and include (1) a school-based apprenticeship approach; (2) school-based teaching research groups; (3) teaching research activities organized by municipality; and (4) teaching contests (Huang, Peng, Wang, & Li, 2010). All these strategies try to make professional development pertinent to teachers' daily instruction, opening it at the same time to public scrutiny. They help to shed light on different aspects of teachers' expertise that are valued as important in developing effective teaching in China.

At a time when China is undertaking curriculum reform, teacher professional development becomes more important. As new topics and innovative instructional ideas are implemented in classrooms, traditional strategies for professional development are not sufficient to meet these new challenges. Recent investigations reveal that professional development has taken the form of adopting new approaches while keeping valuable existing ones (Huang &

Li, 2008). For example, the development of exemplary lessons, the use of hypermedia for case development and distribution, and collaborations with educational experts and teachers from other schools have become new strategies. Nevertheless, one fundamental feature has remained the same: teacher professional development has to maintain the articulation of teachers' needs and establish direct connections to teachers' practices and what teachers try to do in the classroom.

Facilitating Teachers' Professional Development with Practices of Ranking and Promotion

Understanding Teacher Ranking and Promotion in China

Teachers' ranking and promotion practice itself is not a professional development strategy; it functions, rather, as an important mechanism in specifying aspects of teachers' professional expertise valued for promotion and in supporting teachers' professional development. Figure 4.1 highlights the role played by ranking and promotion in providing a supportive context for various professional development activities in China. Without such ranking and promotion as a supportive context, it would be difficult to see these various professional development activities as a coherent and effective whole.

Figure 4.1
Teachers' professional development in China

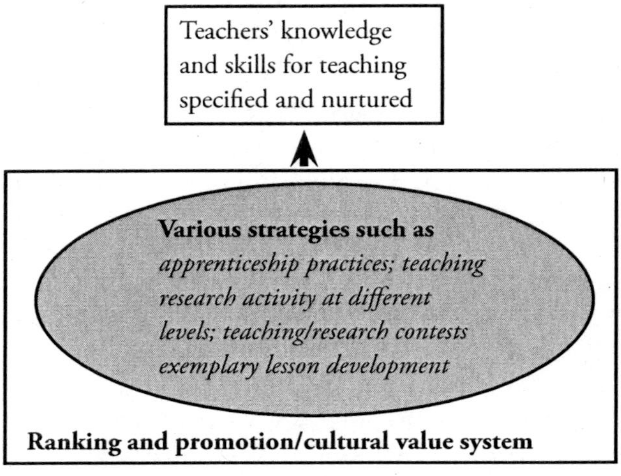

This practice can be further examined from multiple perspectives. In particular, it can be examined from a political perspective to specify how the system is implemented and administrated. It can be considered from an academic perspective to discover what is valorized as part of teachers' expertise as well as from a historical and socio-cultural perspective to understand why this system and this specific conception of expertise work.

From a political perspective, this ranking and promotion system is presented as a bottom-up practice. It is a public policy outlined at the system level but administrated at the provincial level. The system in China functions like the university faculty ranking system in the United States. The rank recognizes a teacher's professional achievement and is linked to his or her salary. However, the rank itself does not guarantee that a senior teacher will have or keep special prestige. In the same way, the system does not prevent junior teachers from moving up; rather, it promotes teachers' professional development.

From an academic perspective, the teacher ranking and promotion system valorizes teachers' professionalism. As the system provides only a general guideline, different provinces promulgate more detailed requirements and methods for ranking and promotion. After comparing requirements specified in different provinces, we found that they share many similarities. For example, an honorary ranking, "teachers of the exceptional class," is given to teachers who are exemplary in several aspects. They are models in terms of displaying morality, educating students, and demonstrating teaching expertise. The general requirements for becoming teachers of the exceptional class include (1) possessing certain qualities, such as having a firm stand on the fundamental mission of the Communist Party, showing patriotism, educating students, and being examples of virtue; (2) holding a senior rank, having systematic and profound theoretical content knowledge, having extensive teaching experience in the subject taught, achieving outstanding results in terms of students' learning outcomes, obtaining outstanding achievements in educational reform, teaching research and curriculum development, and earning a high reputation in the local education field; and (3) making an outstanding contribution in the support of junior teachers (Ministry of Education, 1993). Apart from these general requirements, the provincial regulator (Educational Bureau, Jiansu Province, 2004) provides further details, even including quantitative measures. For example, in specifying teaching quality, it is mentioned that the teacher has developed his or her unique teaching style. The teacher is a leader in the teaching of a certain subject at the municipal or county level and has shown the quality of his or her teaching with some public and exemplary lessons. In terms of the expected contribution to educational reform and research (in particular, teaching method research), the teacher has published a monograph or more than three research papers in journals at the provincial level or at other levels. To illustrate these specifications, we will present, in the next section, the case of an exceptional teacher to make explicit the expertise expected and valued. We will also highlight the teacher's professional promotion during his career.

From a historical and socio-cultural perspective, this system of promotion and its specifications of expertise valorized in China are socio-culturally situated. Because China has a long history of devising fair examination systems for promoting public servants (Li, 2000), it is not surprising that a system of ranking and promotion was used with some specific measures. The practice of specifying and emphasizing the morality and expertise needed for promotion has also been nurtured in Chinese culture. Chinese teachers are respected for having deep knowledge and being moral examples to others. In fact, researchers have documented some cross-cultural differences between China and England over what constitutes a good teacher (Beishuizen et al., 2001; Jin & Cortazzi, 1998). Thus, a historical and socio-cultural perspective is important in developing a better understanding of the Chinese system of ranking and promotion so that it cannot be used directly in another cultural context.

A Case Profile of Teachers' Ranking and Promotion

Teachers of the exceptional class in China are not just experienced teachers. They are also part of the Chinese teaching culture and play an important role in nurturing that culture. To better understand the system of teacher ranking and promotion in China, we will illustrate aspects of expertise that are valorized through the profile of a Chinese mathematics teacher. We will also highlight his trajectory of professional promotion.

Mr. Lin[2] obtained his bachelor's degree in mathematics in 1982 and has been a secondary school mathematics teacher since that time. Due to his great teaching performance, he was transferred two times from one secondary school to another with a better reputation. From 1992 to 2001, he was a mathematics teacher in a key school in an eastern city in China and an associate director of the school's teaching research section. During that period, he completed a master's degree in mathematics and participated in a provincial mathematics teacher leader training program and a national teacher training program. In addition, Lin has published several research papers in key national journals and won several awards in research for the excellence of his publications. His excellence in teaching and research was recognized with the award of excellent junior teacher at the municipal level in 1994. He earned his senior teacher rank in 1996. Those achievements also led Lin to become a mathematics teacher at a key national school in 2001. In the following six years (2001-2006), he made remarkable progress in teaching, research, teacher supervision, and developing instructional materials. He continues to publish research papers in mathematics teaching and develop instructional materials. He presented fifteen exemplary lessons and sixteen teacher training lectures, and he supervised seven junior teachers in an effort to contribute to the implementation of new curriculum. Lin won the title of master teacher at the municipal level and that of exceptional class teacher at the provincial level in 2005. Table 4.1 outlines his professional profile.

Lin described the past twenty years of his teaching career using these characteristics: (1) as a person, he loves students, cherishes life, is a modest and hearty person, has humour, and is honest; (2) as a teacher, he likes to have his own way of thinking, is passionate about education, works hard and is persistent, and has erudition and versatility. Lin realizes that, as a teacher, he pays special individual attention to students, transmits culture through teaching skills, and lays the foundation for their lives and further development.

Based on Table 4.1, we can observe certain relationships between Lin's accomplishments and his changes in position. Before being promoted to a senior teacher, Lin mainly pursued excellence in mathematics teaching. It was this success that led his moves from a local school to a key school at the municipal level. After being promoted as a senior teacher in 1996, he extended his teaching and academic activity to include researching teaching, supervising junior teachers, and participating in educational reform. His excellent performance in these areas led Lin to be honoured as a teacher of the exceptional class. As such a teacher, he has further extended his contributions. Compared with his accomplishments at the senior rank, he has paid more attention to mentoring junior teachers and disseminating his experience in the implementation of new curriculum at the national and provincial levels through textbook writing, public lectures, exemplary lessons, and so on. All this shows that Lin's professional development has been associated with his changes in position.

Lin's professional profile provides us with a picture of different components of his expertise that are valued in China. However, a teacher's expertise

Table 4.1
An outline of Mr. Lin's profile

Period	Award of teaching and research	Publication and project		Supervision	Education reform		
		Paper	Project		Teaching materials	Public lectures	Exemplary lesson
Before 1996.6	Excellent junior teacher (1994, "M") Senior secondary teacher (1996, "P")						
1996.7–2005.6	Mathematics subject leader of intermediate-level teachers (2000, "M") Advanced talent for 21st century (2002, "M") Excellent educational researcher (2004, "M") Master teacher (2005, "M")	2 papers at "N" and 6 papers at "P," 4 of them awarded first-class prizes, the others awarded second-class prizes at "P"	Chair and participant in 6 projects at "M" or above, one of them awarded excellence at "N"	Two teachers	Middle school math textbook (60k words, "N" High school math textbook (400k words, "N")	3 lectures at "P" and 5 lectures at "M"	1 lesson at "P" and 7 lessons at "M"
2005.7–2006.7	Teacher of the exceptional class (2005, "P")			Five teachers	Math textbook at high school (200k words, "N")	1 lecture at "N" and 9 lectures at "P"	1 lesson at "N," 3 lessons at "P," and 3 lessons at "M"

Note: "N," "P," and "M" represent national, provincial, and municipal levels respectively.

is not a simple sum of such components but can often be seen through the teacher's flexible use of expertise in real classroom teaching and his or her views about teaching. A holistic analysis of Lin's expertise is thus necessary to go beyond the list of his credentials.

To develop a better understanding of his master teacher's expertise, we conducted semi-structured interviews with Lin to make explicit his views on effective classroom instruction together with his point of view on two other teachers' videotaped lessons. We also asked five other master teachers to evaluate one of Lin's videotaped daily lessons. Here we first try to clarify his own view on effective teaching and its development, and then we summarize some components of a teacher's expertise that Lin valorized and presented in classroom instruction.

Related to teaching, Lin emphasized "personality and creativity" and the development of a unique teaching style that should attract students. In particular, he developed his own teaching theory and style through the following three stages. In the first stage (1982-1990), as a young teacher, to take account of students' fear of mathematics and their passive learning, Lin attributed their lack of interest to teachers' boring teaching. Then he focused on exploring how to give lessons with passion and revealing the beauty of mathematics. During that period, he gave many exemplary public lessons and was named "the star of public demonstration lesson."

In the second period (1991-1996), Lin shifted his focus from teaching with passion to helping students learn. He realized that in mathematics classrooms a teacher has not only to pursue his individual excellent performance but also, and more importantly, to take care of his students' learning. Then Lin started to explore how students can learn effectively, why they need to learn, and what are the long-term purposes of this learning.

With some deep thinking, explorations, and reflection on these questions, Lin's teaching evolved to the third stage (1997-2006). During the first two stages, Lin focused on the teaching and learning of a certain content knowledge in mathematics and developed a mathematics subject content-centred style of teaching. In the third stage, he started to explore a "student-centred, long-term development oriented" teaching practice. That means that "mathematics is, first of all, a human study." Recently, Lin has explored this teaching approach through the following components.

1. Lin built a student-centred classroom by shifting from the classroom with the teacher as the only principal part to the classroom

with teacher and students as two principal parts. He changed instruction from the teacher-dominated, one-way delivery to that of multiple exchanges between teacher and students, to release students' life spiritualities, to let students laugh in the classroom at least one time in each lesson, and to conceive of the classroom as a place where both students and teacher can learn from each other.

2. Lin changed the traditional approach of practising a high volume of passive problem solving into active problem-solving self-explorations. He also changed short-term imitation exercises into learning by doing long-term projects, and he changed isolated individual problem-solving efforts into multiple exchanges and collaborations. The teacher should have frequent communication with students, commenting on their homework and projects and sharing mathematics and learning experiences with them.

Based on Lin's evaluation of others' classroom instruction and others' views of his classroom instruction, we can highlight some features of the expertise that Lin valorized and exhibited in his classroom instruction:

1. profound understanding of the mathematics subject content in terms of its structure, internal relationships, and development;
2. broad knowledge base regarding historical and contextual aspects of this content;
3. sound curriculum knowledge of content presentation and development over grade levels;
4. overarching of instructional goals that emphasize students' personal development and knowledge growth;
5. teaching design focusing on students' learning by holistic consideration of content structure, curriculum development, and learning difficulties and key points;
6. flexible classroom management through posing worthwhile mathematical problems, questioning, proving, summarizing, and so on; and
7. humour and passion to engage students in the process of learning.

These features, together with the evolution of Lin's views on effective teaching, present a rich picture of his expertise and its development in the context of his professional promotions.

Concluding Remarks

The practice of ranking and promotion, as discussed above, provides a guideline that helps to trace a portrait of teachers' professional development in China. Although different professional development activities/strategies might be available for teachers in different schools, ranking and promotion comprise a wide systemic policy that presents common expectations and criteria for individual evaluations at different rankings. The case study presented in this chapter exemplifies different focuses along the teacher's trajectory of professional development that were consistent with his changes in position.

In many educational systems, evaluating teachers' expertise and promotion have been perplexing. An examination of relevant practices in China can help to advance our understanding of issues and possible approaches. It illustrates a very important aspect of Chinese teachers' professional lives. Moreover, the expectations specified for different rankings and professional promotions in China can also help to explain what is valorized in Chinese classroom instruction. By examining ranking and promotion from multiple perspectives, we can further understand specific aspects of and possible limits to the adoption of such a practice in another system and cultural context. We hope that this chapter will be a catalyst to analyze and understand relevant issues across educational systems and to raise questions to mathematics education researchers regarding a critical analysis of what can be learned from other educational systems.

Notes

1 Master teachers are those who have a senior rank (including an honorary rank).
2 All the names used in this chapter are pseudonyms.

References

Beishuizen, J. J., Hof, E., van Putten, C. M., Bouwmeester, S., & Asscher, J. J. (2001). Students' and teachers' cognition about good teachers. *British Journal of Educational Psychology, 71,* 185-201.

Educational Bureau, Jiansu Province. (2004). *Tentative methods for evaluating and promoting teachers to the exceptional class.* Retrieved 21 February 2008 from http://www.dtjy.com.cn/Article/ShowArticle.asp?ArticleID=91.

Guskey, T. R. (2003). What makes professional development effective? *Phi Delta Kappan, 84,* 748-750.

Huang, R., & Bao, J. (2006). Towards a model for teacher professional development in China: Introducing *keli. Journal of Mathematics Teacher Education, 9,* 279-298.

Huang, R., & Li, Y. (2008). Opportunities and challenges of the effectiveness of in-service mathematics professional development programmes in China. *Journal of Mathematics Education, 17*(3), 32-38.

Huang, R., Peng, S., Wang, L., & Li, Y. (2010). Secondary mathematics teacher professional development in China. In F. K. S. Leung & Y. Li (Eds.), *Reforms and issues in school mathematics in East Asia* (pp. 129-152). Rotterdam: Sense Publishers.

Jin, L., & Cortazzi, M. (1998). Dimensions of dialogue: Large classes in China. *International Journal of Educational Research, 29,* 739-761.

Li, Y. (2000). Purpose, function, and types of mathematics assessment in China: A historical perspective. *ZDM: International Reviews on Mathematical Education, 32,* 192-200.

Li, Y. (2007). Examining and understanding mathematics classroom instruction and possible contributing factors through a cross-national lens. *School Science and Mathematics, 107,* 308-310.

Li, Y., & Huang, R. (2008). Chinese elementary mathematics teachers' knowledge in mathematics and pedagogy for teaching: The case of fraction division. *ZDM: The International Journal on Mathematics Education, 40,* 845-859.

Li, Y., & Kulm, G. (2008). Knowledge and confidence of the pre-service mathematics teacher: The case of fraction division. *ZDM: The International Journal on Mathematics Education, 40,* 833-843.

Li, Y., Kulm, G., & Smith, D. (2006). "Facilitating mathematics teachers' professional development in knowledge and skills for teaching in China and the United States." Invited presentation at the Second International Forum on Teacher Education, 25-27 October 2006, Shanghai.

Li, Y., Ma, Y., & Pang, J. (2008). Mathematical preparation of prospective elementary teachers. In P. Sullivan & T. Wood (Eds.), *Knowledge and beliefs in mathematics teaching and teaching development,* vol. 1 of *International handbook of mathematics teacher education* (pp. 37-62). Rotterdam: Sense Publishers.

Li, Y., Zhao, D., Huang, R., & Ma, Y. (2008). Mathematical preparation of elementary teachers in China: Changes and issues. *Journal of Mathematics Teacher Education, 11,* 417-430.

Ma, L. (1999). *Knowing and teaching elementary mathematics: Teachers' understanding of fundamental mathematics in China and the United States.* Mahwah, NJ: Erlbaum.

Ministry of Education, People's Republic of China. (1993). *Regulations for promoting teachers to the exceptional class.* Retrieved 21 February 2008 from http://www.moe.edu.cn/edoas/website18/info5947.htm.

Ministry of Education, People's Republic of China. (1994). *The Teachers Act in P. R. of China.* Retrieved 16 October 2007 from http://www.moe.edu.cn/edoas/website18/info1428.htm.

Ministry of Education, People's Republic of China. (1995). *Regulations of teacher's qualification.* Retrieved 16 October 2007 from http://www.moe.edu.cn/edoas/website18/info5919.htm.

Stewart, V. (2006). China's modernization plan: What can U.S. learn from China? *Education Week, 25*(28), 48-49.

Part 3

Professional Knowledge Development of Mathematics Teachers

Chapter 5

Secondary Teachers' Professional Development through the Exploration of School Mathematics

Jérôme Proulx

Introduction

In this chapter, I report on a professional development initiative set up for secondary mathematics teachers (see Proulx, 2007). Six teachers (Carole, Carl, Eric, Lana, Linda, and Gina) participated in the project with the intention to revitalize and improve their mathematics teaching practices. The teachers generally thought that their mathematical knowledge was too focused on procedural knowledge and that this orientation had an important impact on their teaching practices. Some of them expressed the following: "Why is it that we are not able to solve by reasoning? ... It is because we have not been educated to reason in mathematics. Me, I did copy, paste, repeat, and let's go, ... and I had 95% in mathematics!" (Carole); "I never understood why it worked. ... When students ask me why, I simply say that this is how it is!" (Lana).

This situation with regard to secondary mathematics teachers is not new, and important critiques have been raised about the mathematical preparation of school teachers. Indeed, studies point to difficulties that secondary teachers have experienced with aspects of the mathematics that they teach. Studies by Ball (1990) and Bryan (1999) have illustrated that, though the secondary mathematics teachers whom they studied made few if any mistakes in their usage of mathematical procedures, they experienced significant difficulties in providing sound meanings and explanations for the mathematical rationales underlying these procedures. Other studies have highlighted difficulties of a different order concerning their unfamiliarity with the meanings of concepts and solving processes (e.g., definitions, conjectures, relationships within concepts). Even (1993) and Hitt-Espinosa (1998) have observed that many teachers possess an "old" definition of a function as a continuous graph, preventing them from recognizing or accepting alternative drawings as representing a function and leading them to transform or treat discrete functions as continuous. Also, Schmidt and Bednarz (1997) and Van Dooren, Verschaffel, and Onghena (2003) have reported on secondary teachers' ease with algebraic solutions but difficulties in appreciating arithmetical procedures as valid solutions to algebraic problems.

Despite criticism that these types of studies present a "deficit model" of teacher knowledge, and are not generalizable to all teachers, they nevertheless offer us significant and insightful information for informing our teacher education practices. The issue appears to be not about what teachers know or do not know but about what these findings tell us about what we could or should offer teachers as mathematical experiences in teacher education initiatives.

Learning from Studies of Teachers' Knowledge: Rethinking Mathematical Experiences

One issue that arises from these studies concerns the significance of investing efforts in teachers' development of mathematical knowledge in teacher education practices. However, as Cooney and Wiegel (2003) explain, one needs to clarify what sorts of mathematics one is talking about. Clearly, the issue is not about academic mathematics,[1] since the underlined difficulties relate to school mathematics knowledge, and academic mathematics simply appears too far removed from the mathematics teaching practices of teachers.[2] The issue therefore appears to concern school mathematics knowledge. This can appear counterintuitive since secondary teachers are considered specialists who know their subject well and thus do not require additional education in mathematics.[3] But the situation is more subtle than a "know/don't know" issue. What we see from these studies is that teachers do know a lot of/about mathematics. Indeed, they have had great success in their student careers and have been teaching mathematics for a number of years. Hence, even if it is reported in some studies that teachers have procedural knowledge or focus too heavily on algebra, this mathematical knowledge is an important part of what they need to know in mathematics. This knowledge can be further developed, as with any knowledge, but does not need to be discarded or erased or seen negatively as "bad knowledge"; secondary teachers need not to "unlearn" what they know but to continue learning about mathematics. Additionally, secondary teachers, for the most part, have a good relationship with mathematics (academically and emotionally). As Cooney and Wiegel (2003) explain, they enjoy doing mathematics and investing themselves in its study. Mathematics is often seen as part of their lives and receives a privileged status: indeed, they decided to pursue a career in teaching mathematics! Thus, mathematics appears as a privileged point of entry to engage secondary teachers in professional development.

In this study, teachers were considered to possess important knowledge, and the aim was to build on and enhance their mathematical knowledge. The content of exploration in the sessions was school mathematics concepts (e.g., algebra, fractions, analytic geometry, volume, etc.), and the activities revolved around offering mathematical tasks for teachers to explore. The specific intention of the research was to study and better understand the learning opportunities that this approach could offer to these teachers.

Methodological Considerations

Ten three-hour sessions were set during the school year, compiling thirty hours of professional development. The sessions, about diverse school mathematical topics chosen with teachers, were videotaped with the camera placed at the back of the room to capture interactions. As the teacher educator, I designed the sessions and chose the tasks to explore (though teachers were often invited to bring tasks on their own). I was also an active participant in the group's discussions and explorations, for tasks served only as starting points for explorations. In short, I did not play the role of a facilitator or guide for the sessions; rather, I engaged myself in exploring the mathematical issues with the teachers.

Tasks were designed on the basis of prior content analyses (didactical and epistemological/historical analyses; see Brousseau, 1998) and of content analyses derived from the research literature (e.g., on algebra, Bednarz, Kieran, & Lee, 1996). In the data reported below, tasks concerned operations with fractions (+, –, ×, ÷), where the focus was on the meanings behind these operations (Kieren, 1976) and the relation to the whole (Hart, 1981).

Exploring Fractions with Teachers: Delving into Algorithms

To illustrate an example of typical work/exploration in the sessions, I report on the seventh session, centred on exploring operations with fractions. This topic was chosen by teachers as they were eager to investigate this topic and to know more about it, largely because they found that their students experienced difficulties with fractions, and they wanted to be able to help them better.

We began by exploring some addition, subtraction, and multiplication of fractions with materials (egg cartons, pieces of paper, and area drawings) and then shifted to division with the folding of pieces of paper and area drawings. To illustrate where teachers were in terms of division of fractions, here is what happened when I offered them $1/3$ divided by $1/4$.

Carole:	Me, I am not able to do this. This, I am blocked.
Jérôme:	And how do you arrive at finding the answer?
Carole:	Me, I know the answer. 1 and $1/3$.
Jérôme:	How do you arrive at it?
Carole:	Technique ["multiplying by the inverse"].
Jérôme:	You do $1/3$ multiplied by 4 over 1?
Carole & Eric:	Yes.

It appeared that teachers knew the algorithm well but had difficulty making sense of the meaning behind the procedure. We therefore aimed to make sense of this operation. Using area drawings, we talked about the division in terms of "how many times $1/4$ enters into $1/3$." So, taking a sheet of paper, $1/3$ represented the following , and $1/4$ represented  or  (and $1/3$ could be represented by ). The operation became how many times 3 squares enter into 4 squares, which resulted in 1 and $1/3$: 3 squares  enter 1 *and* $1/3$ times into 4 squares . The group appeared satisfied with the explanation but wanted to know more about the meaning for the remainder (i.e., the $1/3$), so Carole asked what would be the rest for the division $1/3 \div 1/5$.

To begin this division, we split a grid into 5 and 3, to which Lana, who had been very quiet since the beginning on the division of fractions, objected strongly to this method as we engaged in a conversation on it.

Lana:	Why do you say $1/3$? Why do you divide by 5 first? It is the $1/3$ that is divided by $1/5$. Why do you divide your sheet in fifths?

Jérôme: Because I want to know what $1/5$ is worth.

Lana: No, but, when you read, it is $1/3$ divided by $1/5$. Therefore, I would start. ... I have difficulties; well, I am not able to understand what you are all doing.

Jérôme: How would you do it?

Lana: Well, I start with the first one. We did $1/3 \div 1/4$. I have drawn $1/3$ like this [shows a part of her sheet divided into thirds].

Jérôme: Then. ...

Lana: And then I try to understand what it is.

Jérôme: That is really fine. How many quarters will enter into this [pointing to $1/3$]?

Lana: Yes, but what does $1/4$ represent?

Jérôme: Well, this is exactly it! You have to find out how much does $1/4$ represent. What $1/3$ is worth is this [pointing to the $1/3$ on her sheet].

Lana: Yeah.

Jérôme: And $1/4$, I will try to find what it is worth. Well, if I take the same sheet and I divide it into 4—

Lana: Ha! Ok!

Jérôme: One-quarter is worth this [points to the three squares in the row:].

Lana & Eric: Yes!

Jérôme: Therefore, I am comparing two quantities. How many times *into this* [points to] will enter these ones [points to]? Well, they enter 1 and $1/3$.

Lana: OK, because, OK, OK, OK. There you have your $1/3$ [looking at her sheet], and you tell yourself that you want to divide it by the $1/4$, but what does your $1/4$ represent? ...

Jérôme: That's it, you have to find what your $1/4$ is worth because it is by it that you divide.

Lana: Ok, $1/4$ is 3 squares.

Jérôme: That's it, how many times do 3 squares enter into 4 squares?

Eric: Exactly!

Jérôme: It's 1 and $1/3$.

The discussion continued as we completed the $1/3 \div 1/5$ operation in this way: $1/3$ is 5 squares , and $1/5$ is 3 squares , so 3 squares enter 1 and $2/3$ times into 5 squares . Following this, to step away from uniquely working with unitary fractions, we looked at the following division: $1/3 \div 3/4$. All excited, Lana shouted right away, "$4/9$! There are 4 that fit out of the 9," with Gina being amazed by this reasoning, uttering an excited "Oh!" ($1/3$ is 4 squares  , and $3/4$ is 9 squares  , so 9 squares  enter $4/9$ into 4 squares ).

Teachers were amazed by the depth of the mathematical meanings involved in these operations and how, as Carole said, "a student who understands this will have a level of understanding of fractions that is superior." Teachers therefore started to reflect on the usual algorithm of "invert and multiply" as they knew it and realized that there was much more to it. They wondered why the algorithm worked and directly asked me this: "Why, when I divide $1/3$ by $1/4$, can I invert and multiply?" I therefore felt compelled to offer some ways that could help to make sense of this situation. One of them concerned the context of glasses and litres of water, summarized below.

If I have 6L of water to pour into glasses of 1L, it gives me 6 glasses of water: 6L ÷ 1L glass = 6 glasses.

If now the same 6L of water are poured into $1/2$ L glasses, it gives me 12 glasses, because my glasses are two times smaller, so I need twice as much: 6L ÷ $1/2$ L glass = 12 glasses.

If now I have $1/4$L glasses, it gives me 24 glasses, because the glasses are two times smaller than the previous ones of $1/2$L or four times smaller than the first ones of 1L: 6L ÷ $1/4$L glass = 24 glasses.

If now I have $1/8$L glasses, it gives me 48 glasses: 6L ÷ $1/8$L glass = 48 glasses.

But if now I have glasses of $3/8$L, these glasses are three times bigger than the $1/8$L glasses, so I need three times fewer glasses than with $1/8$L glasses, which is 16 glasses: 6L ÷ $3/8$L glass = 16 glasses.

What came out is that, when we *divide* 6L in $1/2$, $1/4$, $1/8$, or $3/8$, it gives the same thing *as if* we *multiply* by the inverse, namely by 2, 4, or 8 or multiply by 8 and divide by 3 (leading to a multiplication by $8/3$). Not that "it is" multiplied by the inverse but that it ends up giving the same answer. Therefore, $6 \div 3/8$ is not $6 \times 8/3$, but it gives the same answer *as if* we multiplied by $8/3$.

I then gave them another explanation. Since dividing by 1 is far from the simplest division that can be made, if I had to complete a complicated division, for example $^1/_{12} \div {}^3/_4$, I could use the concept of dividing by 1, or organize my operation to obtain a division by 1, by multiplying both numerators and denominators by $^4/_3$:

$$\frac{^5/_{12}}{^3/_4} = \frac{^5/_{12} \times {}^4/_3}{^3/_4 \times {}^4/_3} = \frac{^5/_{12} \times {}^4/_3}{1} = {}^5/_{12} \times {}^4/_3 = {}^{20}/_{36}$$

Lana:	[to Erica] I have never thought about doing that!
Jérôme:	Well, then, it gives me $^{20}/_{36}$.
Lana:	So, in a sense, you can *show* that when you divide by a fraction, it comes up at doing the inverse.
Jérôme:	Yes, exactly, dividing ends up at multiplying by the inverse.

Gina explained to the group that these explanations, in addition to giving meaning to the algorithm of "invert and multiply," made her realize that it was indeed a trick, something she had never seen in these terms before and had mostly seen as a blunt fact. Teachers continued commenting on the reasoning behind the calculations that we had completed as the session came to an end.

Reflecting on Teachers' Knowledge Development

The above is illustrative of comprehensions that teachers explored during the year about school mathematics (here for operations on fractions), concepts that they knew and used frequently, as they worked and reworked in deeper detail these concepts and thus expanded their understanding of them. Davis and Simmt (2006) call this process *recursive elaboration* as one refines and re-elaborates elements of personal knowledge. This is what happened here, where teachers' knowledge was both deepened and refined. In addition, these concepts changed for the teachers. When one compares the meaning that they previously gave to division of fractions and what they appeared to have developed, one sees an important change. Even more, Gina's realization that the algorithm to divide fractions was in fact a trick, something that Gina had never thought of before, made her change significantly her view of the concept. The same can be said of Lana, who commented that she had never thought about ways of showing how and why the algorithm works and offering a sort of proof. In fact, toward the end of the session, both Lana and Eric told me that they had never thought about doing and explaining division of fractions in these ways and would certainly use these explanations in their teaching ("In fact," Eric said, "tomorrow morning!"). Significant here is that these teachers not only gained some mathematical understanding of the division of fractions algorithm but also envisioned possibilities for communicating this understanding to their students and reinvesting it in their teaching. From this we can see an explicit link between their knowledge and their teaching of mathematics. New mathematical understandings appeared to open a space for new teaching possibilities and ways of working with concepts that differ from only offering algorithms (see also Proulx, 2008).

Concluding Remarks and Implications

These mathematical amazements happened continuously throughout the year; teachers explained that the explorations gave them access to ways of understanding mathematics that they had not experienced in their education, ways that were thus unfamiliar to them. As Carole repeated to me and others in the sessions and when I visited her in her school, "I completely changed my teaching!" What strikes me as more important is that the only thing we did was engage in school mathematics concepts, nothing more! It makes the approach focused on school mathematics very promising for teacher education practices, for there is an opportunity to offer mathematical experiences that some teachers simply have not had before.

To conclude, one thing this study makes clear, in addition to pedagogical reflections and the learning of mathematics, is that secondary teachers *can* learn a lot of mathematics. This indicates that, even if research studies point to difficulties or misunderstandings that secondary teachers can have, they *do* have the capacity to learn more mathematics and continue evolving in mathematics. That these explorations opened new mathematical horizons for these teachers is also of significance. As Even (1993) and Hitt-Espinoza (1998) explain, often teachers have not had the opportunity to engage with these sorts of mathematics that go beyond a single usage of procedures. This is encouraging and significant for our practices of mathematics teacher education, in which we have the opportunity to push teachers' mathematical knowledge forward. At a time when concerns are raised about the disconnection between teacher education and teaching practices in school, an approach focused on teachers' knowledge of school mathematics appears relevant since it is directly related to their teaching practices as they develop professional ways of knowing and practising mathematics.

Notes

1 That is, university mathematics courses on advanced mathematics topics (linear algebra, calculus, group theory, number theory, etc.).

2 Concerns are currently raised about the relevance of an advanced study of academic mathematics by secondary teachers since teachers develop professional ways of knowing and practising mathematics strongly disconnected from their teaching of mathematics in schools (Bauersfeld, 1998; Moreira & David, 2005).

3 An assumption that Ball, Lubienski, and Mewborn (2001) and Cooney and Wiegel (2003) question significantly.

References

Ball, D. L. (1990). Prospective elementary and secondary teachers' understanding of division. *Journal for Research in Mathematics Education, 21*(2), 132-144.

Ball, D. L., Lubienski, S. T., & Mewborn, D. S. (2001). Research on teaching mathematics: The unsolved problem of teachers' mathematical knowledge. In V. Richardson (Ed.), *Handbook of research on teaching,* 4th ed. (pp. 433-456). New York: Macmillan.

Bauersfeld, H. (1998). Remarks on the education of elementary teachers. In M. Larochelle, N. Bednarz, & J. Garrison (Eds.), *Constructivism and education* (pp. 213-232). Cambridge, UK: Cambridge University Press.

Bednarz, N., Kieran, C., & Lee, L. (Eds.). (1996). *Approaches to algebra: Perspectives for research and teaching.* Dordrecht: Kluwer.

Brousseau, G. (1998). *Théorie des situations didactiques.* Grenoble: La Pensée Sauvage.

Bryan, T. J. (1999). The conceptual knowledge of preservice secondary mathematics teachers: How well do they know the subject matter they will teach? In *Content knowledge,* vol. 1 of *Issues in the undergraduate mathematics preparation of school teachers.* Retrieved 18 February 2009 from http://www.k-12prep.math.ttu.edu/journal/journal.shtml.

Cooney, T. J., & Wiegel, H. G. (2003). Examining the mathematics in mathematics teacher education. In A. J. Bishop, M. A. Clements, C. Keitel, J. Kilpatrick, & F. K. S. Leung (Eds.), *Second international handbook of mathematics education,* vol. 2 (pp. 795-828). Dordrecht: Kluwer.

Davis, B., & Simmt, E. (2006). Mathematics-for-teaching: An ongoing investigation of the mathematics that teachers (need to) know. *Educational Studies in Mathematics, 61,* 293-319.

Even, R. (1993). Subject-matter knowledge and pedagogical content knowledge: Prospective secondary teachers and the function concept. *Journal for Research in Mathematics Education, 24*(2), 94-116.

Hart, K. M. (1981). *Children's understanding of mathematics: 11-16.* London: John Murray.

Hitt-Espinosa, F. (1998). Systèmes sémiotiques de représentation liés au concept de fonction. *Annales de Didactique et de Sciences Cognitives, 6*(1), 7-26.

Kieren, T. E. (1976). On the mathematical, cognitive, and instructional foundations of rational numbers. In R. Lesh (Ed.), *Number and measurement* (pp. 101-144). Columbus, OH: ERIC.

Moreira, P. C., & David, M. M. M. S. (2005). Mathematics in teacher education versus mathematics in teaching practice: A revealing confrontation. In R. Lins & A. Olimpio (Eds.), *Contributed papers, demonstrations, and worksessions: 15th ICMI study: The professional education and development of teachers of mathematics.* Sao Paolo: ICMI. CD-ROM.

Proulx, J. (2007). *(Enlarging) secondary-level mathematics teachers' mathematical knowledge: An investigation of professional development.* Unpublished doctoral dissertation, University of Alberta, Alberta, Canada.

Proulx, J. (2008). Exploring school mathematics as a source for pedagogic reflections in teacher education. *Canadian Journal of Science, Mathematics and Technology Education, 8*(4), 331-354.

Schmidt, S., & Bednarz, N. (1997). Raisonnements arithmétiques et algébriques dans un contexte de résolution de problèmes : Difficultés rencontrées par les futurs enseignants. *Educational Studies in Mathematics, 32*(2), 127-155.

Van Dooren, W., Verschaffel, L., & Onghena, P. (2003). Pre-service teachers' preferred strategies for solving arithmetic and algebra word problems. *Journal of Mathematics Teacher Education, 6*(1), 27-52.

Chapter 6

Dialogue among In-Service Teachers in an Internet-Based Mathematics Education Program[1]

Mario Sánchez

Introduction

The mathematics education program of the Centro de Investigación en Ciencia Aplicada y Tecnología Avanzada del Instituto Politécnico Nacional (CICATA-IPN) in Mexico is an Internet-based program directed at in-service mathematics teachers in Latin American countries.[2] The program offers master's and PhD studies constituted by a number of different courses. Since the foundation of the program, these courses have been (re)designed based on the experiences, intuitions, and criteria of the different teacher educators in charge of them. In spite of this pragmatic design of the courses, in some cases teachers' engagement in group discussions and reflections positively affect their beliefs and attitudes about the teaching and learning of mathematics.

The aim of the study reported in this chapter was to evaluate, in a more theoretical way, what constitutes these fruitful interactions and the design factors that seem to act in favour of such interactions. The analysis reported here focused on a special design element called *note of reflection,* which has been included in different courses of the aforementioned program. I will describe the characteristics of this element later.

The data presented here were constituted by the interactions among a small group of in-service mathematics teachers while they were discussing one of the notes of reflection. Their interactions were analyzed using the concept of dialogue as it is developed in Alrø and Skovsmose (2002).

Theoretical Framework

The concept of dialogue is recognized as a key element in the learning process (see, e.g., Alrø & Skovsmose, 2002; Chamoso, 2003; McGraw, Lynch, Koc, Budak, & Brown, 2007; Zack & Graves, 2001). An assumption that underlies this concept is that, though the learning is personal, the construction of knowledge is performed through social interactions. In general, the concept of dialogue is understood as interactions in which participants have a genuine interest in the ideas of their interlocutors and intend to discuss and create ideas in a collective way.

In this chapter, I use the concept of dialogue associated with the Inquiry Cooperation Model (IC-Model) of Alrø and Skovsmose (2002). Although the development of the model has been based on the observation of students, collectively solving mathematical open-ended problems, I think that the theoretical structure is sufficiently general to be used in the analysis of computer-mediated interactions among in-service mathematics teachers. The model, strongly rooted in the critical mathematics education approach, argues that a fruitful interaction must be based on mutual respect, willingness to make public our ideas and subject them to scrutiny, and a real interest in listening to and analyzing our interlocutor's ideas.

The IC-Model is constituted by a set of communicative characteristics. According to this theoretical approach, an interaction, like the one previously described, should have several of these communicative characteristics. In fact, when these characteristics are present in an interaction, it is regarded as a special kind of interaction called *dialogue,* which can serve as a basis for critical learning and reflection. The communicative characteristics that define a dialogue are *getting in contact, locating, identifying, advocating, thinking aloud, reformulating, challenging,* and *evaluating,* and they can be defined as follows:

> *Getting in contact* involves inquiring questions, paying attention, tag questions, mutual confirmation, support and humour. *Locating* has been specified with the clues of inquiring, wondering, widening and clarifying questions, zooming in, check-questions, examining possibilities and hypothetical questions. *Identifying* involves questions of explanation and justification and crystallizing mathematical ideas. *Advocating* is crucial to the particular trying out of possible justifications, and it is closely related to arguing and considering. *Thinking aloud* often occurs as hypothetical questions and expression of thoughts and feelings. *Reformulating* can

> occur as paraphrasing, completing of utterances and staying in contact. *Challenging* can be made through hypothetical questions, examining new possibilities, clarifying perspectives, and it can be a turning point of investigation. *Evaluating* implies constructive feedback, support and critique. (Alrø and Skovsmose, 2002, p. 110)

I am particularly interested in the study of interactions among in-service mathematics teachers because I advocate that it is important for them to be involved in dialogical interactions. Generally, in-service mathematics teachers possess a collection of experiences, knowledge, and beliefs that influence their teaching practice. A positive way to achieve changes in those beliefs, to identify their conceptions, and to sensitize them on the complexity of the mathematics learning process is through the expression, comparison, debate, and argument of their own ideas and points of view. It is where a dialogue can serve as the basis to achieve critical reflection on those ideas, knowledge, and beliefs.

Methodology

In this section, I will refer to three aspects of the methodology used in the study; specifically, I will give a description of the mathematical activity used (the so-called note of reflection), I will discuss the course where the note of reflection was embedded, and finally I will briefly describe the data collection and presentation.

The Mathematical Activity (Note of Reflection)

A note of reflection is a written case describing a fictitious classroom event arranged around a mathematical question or activity. In this particular study, the event takes place in an imaginary setting in which a teacher asks her students a question. Although the event did not happen in reality, the answers provided by the imaginary students (called student A and student B) have been inspired by real teaching experiences or have been taken from experimental data included in some regional research thesis or report.

After looking at the situation, teachers are asked to explain the possible causes behind the answers of students A and B. Then they are asked to decide if they consider the answers to be mathematically correct. Figure 6.1 shows an English translation of the note of reflection presented to the teachers.

Figure 6.1

Analysis of the mathematics school discourse II

SITUATION

A mathematics teacher asks the following question to their students:

The equality 0.99999999... = 1 is it true?

Two of their students give an answer to the question. These are the answers and the arguments that they used:

Student A
I think that 0.99999... is not equal to 1, because although 0.99999... is very close to 1, always is going to lack a small quantity to be equal to 1

Student B
I think that 0.99999... = 1 is true because if I use a technique to convert a decimal number to its fractional expression, that I have that:

$x = 0.999999...$ (1)

Multiplying for 10 both side of the equality we obtain:

$10x = 9.999999...$ (2)

Carrying out the difference among (2) and (1) we have that:

$$\begin{array}{r} 10x = 9.999999... \\ x = 0.999999... \\ \hline 9x = 9 \end{array}$$

Then $x = 1$

QUESTIONS

1. Which do you think that are the ideas that produce the answer of student A?
2. Which do you think that are the ideas that product the answer of student B?
3. In your opinion, which student is correct in their answer? Why?

The note of reflection is a useful tool for the structure of an in-service course for teachers. Its "fictitious" character allows the teacher educator to create specific situations or cases anchoring teachers' attention and discussion in particular issues or situations that in turn correspond to the aims of the course or the particular interest of the designer. In the next section, I will briefly describe the course in which the note of reflection was implemented.

A Brief Description of the Course

The course in which the note of reflection was applied was called Analysis of the Mathematical School Discourse II. It was included in the curriculum of the master's studies in mathematics education of the CICATA-IPN program and lasted four weeks. It was taught during 2005, and twenty-seven teachers from three different Latin American countries (Argentina, Uruguay, and Mexico) attended it. The participant teachers were working at different educational levels (elementary, upper secondary, and university level).

The course focused on the study of different cognitive theories in mathematics education research. The main goal of the course was to enable teachers to use cognitive theories to analyze their students' behaviours and to be aware of their own mathematical conceptions of some mathematical concepts, such as that of infinity.

The course was composed of different activities, such as reading mathematics education research papers; watching a video containing the application of a didactic activity aimed at producing a cognitive conflict; writing the previously described note of reflection; and engaging in group discussions carried out in Internet-based forums by means of asynchronous exchanges of written messages.

The data that I will present correspond to an interaction among a small group of teachers discussing the contents of the note of reflection presented in Figure 6.1. I will explain how the data were collected and how they will be presented in this report.

Data Collection and Presentation

Each note of reflection was answered by the teachers individually and delivered to the teacher educators in charge of the course as a word file through the Internet-based workplace. The responses were reviewed with the aims of identifying teachers with different ideas about the mathematical content of the note of reflection and putting them into the same working group. In other words, heterogeneous working teams were constituted where some of the team members thought that $0.9999 \ldots = 1$ is true, while others had different ideas about it. The initiative of setting up heterogeneous groups responded to theoretical arguments claiming that this is a design element that favours a computer-mediated dialogue (see, e.g., de Vries, Lund, & Baker, 2002; McGraw et al., 2007).

One characteristic of the computer-mediated communication is that it can easily be recorded, stored, and shared. This feature represents a significant advantage for educational research because the need to make transcriptions disappears. In this work, for instance, where I analyzed some of the written asynchronous discussions produced during a course taught in 2005, those discussions are permanently recorded and accessible on the Internet-based workspace, ready to be analyzed.

These asynchronous discussions usually lasted several days and could be composed of dozens of utterances. Due to the space available, it is not possible to present the complete interactions, only those sections that I consider the most significant and illustrative. The data that I will present have been translated from Spanish into English.

Results

In the following excerpt, an interaction takes place between two teachers called Ana and Paula.[3] Ana is Mexican, and Paula comes from Argentina. Ana works in secondary-level education. She has a bachelor's degree in mathematics and works on the production of educational videos for television-based distance education. Paula studied to become a mathematics teacher and has three workplaces: she works as a teacher in an elementary school and a secondary school, but she is also a teacher educator in a teacher education institution in Argentina.

The discussion begins with some thinking aloud in which each teacher explains her ideas about the students' answers. Paula says that, though the expressions 0.9999 ... and 1 are equivalent, the method used by student B is not applicable to decimal numbers with infinite expansion. Ana, in turn, thinks that the responses of both students might be correct (see [2]).

> [1] *Date: 17-11-2005 08:55*
> *Paula:* I am still thinking on 0.9999 ... and 1. I would like to know how this topic is presented in Mexico. In Argentina, we have a very structured and not so visual training. When the first note of reflection was presented, I did not doubt it, the answer was that they [these numbers] are

equal. Nevertheless, the things that have been said in this forum, and in the previous one, cause me to have some doubts regarding the way you teach this aspect of a decimal representation. Can you tell me, how do you teach the relationship between fractions and decimals, and how do you define the number 0.9999 ... as a real number? Thanks.

[2] *Date: 17-11-2005 11:44*
Ana: The truth is that I have not tried the topic with students (I do not work in front of a group), but I am an adviser for a didactic guide that will include this topic, and for this reason I am interested in this reflection. When I was a student, I saw it in two ways (separately, as it is commonly studied). For that reason, it seems to me that both are correct in the context in which they are used. If somebody has worked this question in a classroom, I hope you can share your experience with us.

[3] *Date: 17-11-2005 11:46*
Paula: I do not understand what you mean when you say that in one context the first answer is correct. I do not find that context inside the mathematics I know. Perhaps I don't know it, and there is the source of my lack of answers in the discussion on the first forum. In what context is A correct? Can you explain it to me?

Ana is *getting in contact* with Paula in [2]. Ana explains that she has no experience teaching this topic, but she explicitly states her interest in the discussion. She refers to her experience as a student to express her idea of the validity of the responses of both students. This provokes a reaction by Paula, and in [3] she is trying to locate Ana's perspective on the validity of the answers. Paula expresses her lack of understanding of Ana's idea, but at the same time she is willing to examine her own understanding: "Perhaps I don't know it, and there is the source of my lack of answers in the discussion on the first forum." This sort of attitude is a clear indication that we are hearing a dialogue. As Alrø and Skovsmose (2002, p. 121) affirm, "a dialogue also means *willingness to suspend one's perspectives*—at least for a moment." Ana responds to the questions from Paula, bringing to the discussion the mathematical concept of density.

[4] *Date: 17-11-2005 12:13*
Ana: I am talking about the idea that the student is using to give his answer, in this case the density of the real numbers.

[5] *Date: 17-11-2005 12:18*
Paula: The problem that I see is that it has nothing to do with density, the problem is that they are the same number; it is not that they are two very close numbers. They are the same one. It is there where the student gets confused. It is the way I see it.

[6] *Date: 17-11-2005 12:26*
Paula: The problem is that I cannot have two different answers for the same question. In the video we saw, Janet finds two different answers, both correct from her point of view. But what finally is concluded is that it is not possible, within the mathematics body, that two different answers for the same question coexist. If this could be possible, then all could depend on the ideas from everyone. Definitions exist to avoid that.

Discussion

The note of reflection was an element that led to the emergence of different perspectives on the validity of 0.999 ... = 1, which in turn functioned as the basis for the interactions between Ana and Paula. The analysis shows that this note of reflection seemed to be a decisive factor in the initiation of communicative acts that can be considered dialogues.

Other factors also seem to favour the emergence of a dialogue. As de Vries, Lund, and Baker (2002) and McGraw et al. (2007) observe, we recognize that the emergence of a dialogue is favoured among heterogeneous groups, as was the case with Ana and Paula (members of a group with different academic backgrounds and/or different ideas on the same topic). However, heterogeneity is a condition that supports dialogue but does not guarantee its occurrence. For instance, other teachers had different interpretations of the equality presented in the note of reflection, and they were online at the same time as Ana and Paula were interacting, but they were not involved in the discussion in an active way. Why did Ana and Paula engage in a dialogue? A theoretical explanation can be found in the concept of intention of Alrø and Skovsmose (2002).

Participation in a dialogue can be determined by the background and foreground intentions of the participants. Alrø and Skovsmose (2002, p. 160) affirm that "we must consider how the activities in the classroom relate to their [the students'] priorities that are constructed by their hopes, expectations and frustrations when they face the future. ...

Foreground refers to the prospects of the person." Thus, excerpt [2] shows that Ana is interested in the discussion, because the content is related to the professional responsibilities that she will face in the future. Then, to produce teachers' engagement, the proposed activities should not be only open activities that accept different solutions aiming to produce a variety of perspectives. Another important condition is that those tasks should be significant for teachers: that is, they should be related to their background and foreground intentions.

My interest in the creation of situations and the identification of conditions that favour dialogue among teachers resides in my conception of dialogue as an act intimately connected to reflection. The dialogic acts (e.g., getting in contact, thinking aloud, etc.) present in the interaction between Ana and Paula allow me to affirm that they were involved in fruitful interaction or *dialogue:* that is, a rich interaction in which teachers express, justify, compare, and evaluate their own mathematical and pedagogical ideas. This dialogue is a source of reflection; for instance, at the end of the course, we asked Paula if it had contributed to her professional practice, and she answered that "through the forums I am able to know the way of thinking of some of my colleagues and find the similarities and differences between our ideas. And I understand also that all the different views we have on the same reality are reflected in our students." This suggests that Paula understood that teachers' ways of thinking and how contents are presented can influence the mathematical development of students.

Finally, I would like to address how the data presented in this chapter elaborate the theoretical tool applied. First, the data are evidence that application of the IC-Model can be extended to analysis of human interactions in online settings. Nevertheless, it would be necessary to continue studying how the communicative characteristics that define the IC-Model are influenced and shaped by the technological tools used in an Internet-based educational setting. An example is found in [6], where Paula uses the content of a video to *evaluate* Ana's idea of the existence of two different answers for the same mathematical question. Second, the IC-Model seems to be a suitable theoretical tool to investigate how the communicative characteristics present in an interaction influence teachers' reflections and thoughts; however, it seems to me that the tool does not capture the influence of non-human elements from an online setting in the development of reflections and ideas. I consider this aspect a major limitation of the tool, especially if somebody wants to study the influence of the characteristics and tools of the online setting in the professional development of mathematics teachers.

Notes

1 This work was supported by the Programme Alβan, the European Union Programme of High Level Scholarships for Latin America, scholarship number E06D101377MX.

2 More information (in Spanish) about the program can be found at www.matedu.cicata.ipn.mx.

3 The original names have been replaced to protect teachers' identities.

References

Alrø, H., & Skovsmose, O. (2002). *Dialogue and learning in mathematics education: Intention, reflection, critique.* Dordrecht: Kluwer.

Chamoso, J. M. (2003). Considering dialogue as a social instrument in the mathematics class. *For the Learning of Mathematics, 23*(1), 30-40.

de Vries, E., Lund, K. & Baker, M. (2002). Computer-mediated epistemic dialogue: Explanation and argumentation as vehicles for understanding scientific notions. *Journal of the Learning Sciences, 11*(1), 63-103. doi: 10.1207/S15327809JLS1101_3.

McGraw, R., Lynch, K., Koc, Y., Budak, A., & Brown, C. A. (2007). The multimedia case as a tool for professional development: An analysis of online and face-to-face interaction among mathematics pre-service teachers, in-service teachers, mathematicians, and mathematics teacher educators. *Journal of Mathematics Teacher Education, 10*(2), 95-121. doi: 10.1007/s10857-007-9030-3.

Zack, V., & Graves, B. (2001). Making mathematical meaning through dialogue: "Once you think of it, the z minus three seems pretty weird." *Educational Studies in Mathematics, 46*(1–3), 229-271. doi: 10.1023/A:1014045408753.

Chapter 7

In-Service Professional Development for Teachers of Pre-University Mathematics

Bernard Murphy

Background

There is a well-publicized shortage of specialist teachers of mathematics in England (see, e.g., Advisory Committee on Mathematics Education [ACME], 2002; Department for Children, Schools, and Family [DfES/DTI], 2006; Smith, 2004). In response to the Smith Report, the UK government set up the National Centre for Excellence in the Teaching of Mathematics (NCETM) to coordinate the national provision of professional development.

Both the Smith Report and the NCETM highlighted a gap in the professional development available to mathematics teachers. They highlighted the need for professional development courses that provide opportunities to secondary school mathematics teachers to extend their subject knowledge and deepen their understanding of the associated pedagogy.

In 2003, Mathematics in Education and Industry (MEI), an independent UK curriculum development body with many years of experience in providing professional development, designed the Teaching Advanced Mathematics (TAM) course in collaboration with the University of Warwick. TAM is aimed at teachers who are competent to teach students up to age sixteen, to the General Certificate of Secondary Education (GCSE) qualification in England, and who wish to develop their knowledge, skills, and understanding of A-level mathematics, and its teaching (see Porkess, 2003, for information related to the structure of A-level mathematics). In June 2004, with the funding support of the Gatsby Charitable Foundation, a pilot phase was initiated with twenty teachers enrolled in the course. Due to numerous demands, the course is now offered through four universities in England, supporting approximately seventy teachers per year. Successful completion of the TAM course leads to a postgraduate certificate in A-level mathematics pedagogy, which constitutes one-third of a master's degree.

Underlying Principles Guiding Design of the Course

In 2002, leading practitioners and experts in the field of mathematics education in England met to reflect on principles guiding the design of high-quality professional development opportunities for mathematics teachers. These principles were reported by the Advisory Committee on Mathematics Education (ACME, 2002) and informed design of the TAM course, as delineated below.

- Promoting collaboration between different institutions: TAM is based on collaboration between MEI, universities, and schools.
- Availability of professional development: TAM aims to support teachers (with the necessary experience) who want to develop their teaching to a higher level.
- Availability of non-teaching time to reflect on teaching practices: TAM course days give participants time away from the pressures of the classroom.
- Sustaining continuous professional development: TAM is long term, lasting sixteen months, with three years' support in total.
- Teachers' ownership: TAM has built-in flexibility, giving teachers the opportunity to choose areas of study outside a common core.
- Independence: funding from the Gatsby Charitable Foundation allowed MEI to design TAM without externally imposed constraints.
- Deepening subject knowledge: TAM assumes knowledge of the General Certificate of Secondary Education and allows teachers to deepen their knowledge to A level and, if required, beyond.

The course was based on these well-established, effective principles. Its development was also based on an evaluation process.

Integration of Assessment Processes in the Course Design

During the pilot phase, course development was influenced by the findings of external evaluators from the Curriculum, Evaluation, and Management Centre (CEM) at Durham University. These researchers adopted a mixed methods research approach, including questionnaires, interviews with participants, interviews with teaching colleagues and students, observations of lessons, and observations of university-based sessions. They followed participants through the course and monitored their progress for one year after completing the course.

This evaluation was an integral part of the pilot phase. The researchers produced regular reports as the pilot phase progressed for the TAM Steering Committee (made up of members of MEI, the Gatsby Charitable Foundation, and the four participating universities). The assessment study was a formative one, offering suggestions and recommendations for course improvement responded to by the course manager and university providers. In the six interim reports (between 2005 and 2007), there is much evidence from the participants that they got a great deal from it in terms of building their knowledge and confidence to teach mathematics to A-level mathematics students. (This was also supported in the final report of CEM, presented in August 2008.)

In 2008, participants who had completed the course were asked in an email survey to give their views on how the course continued to influence their teaching along with details of any other continuous professional development they had subsequently undertaken. Their unedited responses can be found in the report "A Gateway to Teaching Advanced Mathematics."[1] Further research is currently being undertaken by the NCETM as part of their Researching Effective Continuing Professional Development in Mathematics Education (RECME) project and by Jill Adler through the use of the QUANTUM research methods (see Adler & Davis, 2006).

The Course Structure

I will briefly characterize the structure and main features of the course. TAM is a sixteen-month course that runs from June one year to September of the following year, thereby incorporating two summer holidays, each approximately six weeks in duration. This structure gives time for participants to work without being interrupted for a substantial period of time away from the pressure of the school environment (researching assignments and studying mathematics), enabling them to focus on the course requirements.

Approximately nine days are spent at the university, focusing on the subject content of A-level mathematics, its pedagogy, and related issues such as student misconceptions, assessment, information and communications technology (ICT), the role of proof in mathematics, and mathematical modelling. Participants are challenged to think about mathematics in a way that takes them beyond the comfort zone of typical A-level textbooks. Course leaders facilitate discussion on aspects of pedagogy such as the use of questioning, student-centred activities, and differentiation and set associated assignments in the form of essays and classroom-based research, leading to the award of a postgraduate certificate.

The course requirements involve a substantial amount of private study. This is time consuming for participants, but the evaluation reports that most participants respond positively to this issue and appear to benefit from the personal development involved in learning to manage effectively their study time in relation to family- and work-related commitments.

Extensive support is provided to help participants learn the content of the four core mathematics modules and their chosen areas of applied mathematics. This support is provided through drop-in sessions at the university; by web-based resources in the form of teaching materials such as lesson plans, interactive ICT resources, active learning resources, and enrichment materials; by email; and through weekly online lessons using the Elluminate virtual classroom.

Throughout the course, all participants teach A-level mathematics to sixteen– to nineteen-year-old students and receive at least two supportive school visits. Participants are observed teaching their students in their own classrooms, with the lesson videotaped, and then they receive feedback on how their preparation and delivery of a lesson might be improved.

Throughout the course, participants benefit from the three-pronged collaborative support of MEI, of their university, and of their school/college. Specifically, TAM is centrally managed by MEI through the appointed course manager, who brings coherence to the course at the various universities around the country at which it is offered, to the benefit of all participants. MEI ensures that high-quality online resources are available to participants at all centres to support their learning and teaching of A-level mathematics. In addition, MEI provides participants with all the support they need in terms of acquiring subject knowledge, understanding its place within the wider framework of mathematics, and experiencing a range of effective ways of teaching it. The purpose-written, web-based resources, online lessons in a virtual classroom, and regular emails give teachers focus in and structure to their independent studies. The school visits, undertaken by MEI and the university course tutors, give participants one-to-one advice, support, and ideas for use in their own classrooms.

The university introduces participants to the research underpinning the teaching and learning of A-level mathematics. The university-based study days allow MEI and the course leaders at the universities to model good practice, to discuss the major issues related to teaching at this level, and to provide any other face-to-face support as required by individual participants. The university designs the course assessment, accredits the work at the postgraduate level, and provides further options for participants to continue their studies at Master's level for those who wish to complete a higher degree qualification.

Schools and colleges support the participants by providing them with the teaching experience required to put into practice the skills that they are gaining from the course and by releasing them for the university-based days. When it is possible, colleagues with experience in teaching A-level mathematics provide daily face-to-face support as required. In this way, the course provides many of the participants with new directions in their teaching careers and is a gateway to further professional development. This is evidenced in the six interim reports produced so far by the external evaluators.[2]

Patterns of Delivery, Evolution in the Course Approach

During the pilot phase, TAM continually evolved. Improvements arose as a consequence of the collaborative work between the universities and MEI and in response to the feedback from participants. In particular, the independent feedback provided by the evaluators, through interviews with course participants, resulted in specific improvements to the course. They included redesigning the assessment to make it more accessible to participants, developments in the ways that participants are supported both face to face and online, and changes in how participants study applications of mathematics (this issue includes using national experts as consultants to deliver specific university study days).

Although all four universities have broadly similar modes of delivery, the universities themselves are independent awarding bodies whose requirements for accreditation and certification differ, so local courses must meet the requirements of the university at which they are based. The flexibility in the program has allowed each university to design a distinct way of assessing the course. This has included essays on aspects of A-level mathematics pedagogy, "classroom-based enquiry" involving research into an issue of the participant's choice, and mathematical investigation in which participants reflect on the process of working on mathematics and the implications that it has for their teaching.

Lessons Learned from the Pilot Phase

In designing the program, we aimed to answer two questions:

- Given the restrictions on teachers' time, is it viable to run a sustained course of professional development to equip non-specialist teachers with the skills to teach A-level mathematics with confidence?
- Can a sixteen-month course set teachers on a route of continuing professional development giving them the independence to acquire further knowledge and pedagogical skills?

The evidence from the course participants, their colleagues, the course evaluators, and the course providers clearly indicates that the answer to both questions is "yes" and that the existing course, as previously described, is highly successful in achieving its aims. I will articulate some of the results from this study.

The motivation for enrolling on the course differs between participants, but for all the most important aspect is to develop their knowledge of mathematics and teaching skills to enable them to teach at A level.

According to the evaluation team, most participants see the university assessment as relevant to their professional lives and thus choose to complete the assignments required to qualify for the postgraduate certificate, even when this was not their intention at the outset. For some participants, the qualification is seen as an important aid in seeking promotion, whereas for others this is not an issue.

It has been important to get the balance right on the university study days between devoting time to mathematics and its pedagogy while also ensuring that the university's requirements for what constitutes an academic course at this level are met. Feedback from the participants as evidenced in the evaluation reports clearly indicates that the balance achieved meets their expectations and needs.

Mixtures of after-school, half-day, full-day, and Friday-Saturday university tuition sessions have been trialled, and the latter have been by far the most successful. They are now used at all the universities. In responding to the questionnaires from the independent evaluators, participants describe themselves as benefiting from the more intense focus and believe that the compromise between their own time and school time is appropriate. Nevertheless, many prospective course participants report that their schools are not prepared to release them for university study days in term time due to the associated costs of providing supply teachers, and consequently they cannot enrol on the course. In the proposed expansion, from four universities in the pilot phase to ten participating universities throughout England, we are recommending that the schools and colleges of participants receive funding to meet these costs and thus recognize their contributions in tackling the national shortage of well-qualified mathematics teachers.

Some Course Outcomes

While on the course, participants develop their understanding of mathematics to a much greater depth. This has a positive effect on their teaching at all levels from typically age eleven upward, thus enabling them to present a joined-up view of mathematics throughout a student's time in secondary school. The participants develop a greater appreciation of mathematics pedagogy and acquire the motivation to continue their professional development in mathematics education beyond the completion of TAM.

The resources and pedagogy that participants are introduced to in the course are shared with their teaching colleagues in their schools and colleges, and as a result there is a substantial improvement in the general standard of mathematics teaching at all levels.

Many of the participants were originally trained in specialties other than mathematics in which there is no shortage of teachers. In changing careers to become mathematics teachers, capable of teaching to A level, they represent an additional, and cost-efficient, way of increasing the number of well-qualified mathematics teachers, thus making a significant contribution to the supply of mathematics teachers nationally. It is planned for TAM to expand and be offered at ten universities in England, with about 200 participants a year. It is believed that the TAM model can be adapted to help the supply of teachers of mathematics in other countries where there is a shortage.

Conclusion

The pilot course showed that TAM can be an appropriate model for the professional development of teachers of mathematics. Central management of the course through MEI has proven crucial in ensuring that high-quality, but economically viable, resources are available to all participants both online and during study days and that "good practice" is shared between the course providers. The course benefits from this collaborative expertise offered by universities, schools and colleges, and MEI. The university provides up-to-date knowledge of research on pedagogy and has the authority to award postgraduate qualifications. MEI provides skilled classroom teachers and examiners to support participants and expertise in producing online resources. This collaboration has proven highly effective in meeting the needs of teachers. The role of the schools is also important as they allow teachers to teach A-level mathematics and support them. The independent evaluation by a group of researchers (from CEM) highlighted the potential of this course. However, difficulties should also be mentioned. One of them is the lack of funding for supply teachers so that schools can release teachers for university study days. It seems important that the money going into schools for professional development be ring-fenced so that teachers can access this support. Clearly, many teachers wish to do so.

Notes

1 See http://www.mei.org.uk/files/pdf/TAM_Report.pdf.

2 These reports can be seen at http://www.mei.org.uk/files/pdf/TAM_Report.pdf.

References

ACME. (2002). *Continuing professional development for teachers of mathematics.* Retrieved 19 February 2009 from http://www.acme-uk.org/event.asp?id=55&archive=true.

Adler, J., & Davis, Z. (2006). Opening another black box: Researching mathematics for teaching in mathematics teacher education. *Journal for Research in Mathematics Education, 36*(4), 270-296.

DfES/DTI. (2006). *The Science, Technology, Engineering, and Mathematics (STEM) Programme report.* Retrieved 19 February 2009 from http://www.dfes.gov.uk/hegateway/uploads/STEM%20Programme%20Report.pdf.

Porkess, R. (2003). The new AS and A-levels in mathematics. *MSOR Connections, 3*(4), 12-16. Retrieved 19 February 2009 from http://www.mathstore.ac.uk/newsletter/nov2003/pdf/asa.pdf.

Smith, A. (2004). *Making mathematics count: The report of Professor Adrian Smith's inquiry into post-14 mathematics education.* London: Stationery Office.

Part 4

Experiences of Professional Development in Teaching Mathematics

Chapter 8

In-Service Teacher Education: A Collaborative Experience with Non-Specialist Teachers Who Teach Mathematics[1]

Nielce Meneguelo Lobo da Costa

Introduction

This chapter reports on research on an in-service teacher education program involving primary-level educators exploring mathematical and statistical contents using computers. The research was part of a broader study whose aim was to identify the characteristics of a pedagogical educational process that contributes to the "development" of teaching practice. Development here is understood as the way in which teachers reorganize—or adapt—their didactic practices through reflection, action, and new reflection on their teaching practices. Therefore, the research intended to answer the following question: what are the main factors that contribute to the development of teaching practice in an in-service teacher education process using computers as a tool in the construction of learning situations?

The investigation considers the educational process of teachers as a continuous learning process throughout their lifetimes. In examining the several forms of teacher education programs and their paradigms, we can notice that they are not autonomous; as Pérez Gómez (1997) says, they are in fact contextualized, and, among other things, their guidelines depend on how they conceive of education and curriculum. This means that there are different ways of understanding educational practices related to various forms of teacher education programs.

The educational program model chosen is that of "practical reflection for social reconstruction" (Pérez Gómez, 1997), in which the reflective practice concept is enlarged to consider, besides the reflection in action and the reflection on action, in the sense of Schön (1983), the reflection on the reconstruction of situations in which the action is produced, so that teachers can reconstruct for themselves (as teachers) the situations and the underlying educational assumptions. In this model, there are two different theoretical perspectives. The first one refers to programs that advocate political engagement to promote equality, justice, and social emancipation. The second one, used in this research, refers to *action-based investigation and comprehension-intended,* which embraces the idea that the process of in-service teacher education is oriented by democratic methods and principles, without seeing the teacher acting as a political subject. In this chapter, I will focus on the collaborative experience of a group of teachers engaged in the latter perspective.

The Research and the In-Service Teacher Education Program

A work team called Action Group formed by four university researchers[2] and five teachers in a public school in São Paulo State, Brazil, was part of an in-service educational program held in the school. The research actors were the principal, Hydrangea; the pedagogical coordinator, Violet; and three ordinary primary teachers, Rose, Margaret, and Orchid,[3] all engaged as teacher educators for all the school teachers.

The Action Group met over twenty months on a weekly basis. The sessions focused on the creation of materials for the classroom, on learning situations for students and teachers, and on the elaboration and realization of some workshops with pupils and other teachers. They also involved joint reflection on the learning process. During these meetings and throughout the whole project, members of the Action Group dealt with mathematical concepts and main statistical principles while analyzing computer databases specifically developed for the research actors' context.

Qualitative research characterized as co-generative action-based investigation among collaborative groups was developed in this in-service educational approach. Greenwood and Levin (2000, p. 96) define *co-generative action-based investigation* as research built on professional researcher-stakeholders' collaboration and aimed at developing new knowledge. This type of action research is grounded in the following characteristics:

1. Participants and researchers co-generate knowledge through collaborative communicative processes in which all participants' contributions are taken seriously. The inquiry process leads to the construction of new meanings.
2. It treats the diversity of experiences and capacities within the local group as an opportunity for the enrichment of the research process.
3. It produces valid research results.
4. It is rooted in context, centred on this context, and aims to solve real-life problems.

The in-service teacher education program was strategically divided into four "phases" to facilitate the analyses, as described below.

- Phase 1 was held in the university, and the first meetings discussed the software to be used and planned the in-service educational program to be developed for school teachers.
- Phase 2 occurred in the school during the first semester as an initial hands-on period, when the in-service educational program started. At that moment, two different groups were formed: the Action Group and the Hands-On Group. In this phase, the Action Group developed workshops with students (inside and outside the IT room), came up with preparatory works for a presentation in an educational congress, and realized the investigative studies on their pedagogical practices as part of the project.
- Phase 3 happened during the second semester in the school as the most important part of the project. The Hands-On Group started to create learning situations for pupils. At that point, they started to believe in and really accept the project. As time went on, for the Action Group, the acting sphere increased. Beyond conceiving and developing tasks for teachers, they started to follow and give support to the hands-on teachers when they worked with pupils. At that moment, it was already possible for each member of the Action Group to reflect on his or her own action as teacher educator as well as on his or her work with other teachers when supporting them in their pedagogical practice.
- Phase 4 took place in the last months of the project, during the next academic year, when the Action Group could no longer count on the school's pedagogical coordinator and on the principal as

members of the team.[4] With the new principal, the in-service educational program restarted under new political and practical conditions. For the Action Group's teachers, it was not only a period of concept revision and deep studies but also a return to IT laboratory classes with students and school teachers. At that moment, we, from the university, noticed that they became much more autonomous and that they considered the project as their own. As a last activity, the Action Group took part in a scientific meeting in which the members gave a presentation on the project.

Analysis and Outcome

The research data analysis was qualitative. A triangulation of data (Mathison, 1988) was conducted using four educative dimensions experienced by the research actors. These dimensions were created as categories of analysis to deeply understand each role that they played during the project: student, teacher (with students), teacher educator, and researcher.

Each role taken in the group, as we will see below, was tied to a unique contribution to professional growth. I will consider more particularly the teacher educator role because it seems to be the most significant in promoting cooperation among the Action Group members.

- *Student role:* Its importance can be attached to the need for "reconstruction" of the meaning of teaching mathematics. It can also be related to the need for change in the members' view of mathematical learning among children. Throughout the educational process, members not only constructed another meaning of the mathematical content to teach but also developed new knowledge, creating data-handling strategies, setting up students' information technology (IT) abilities, and, finally, increasing the use of Tabletop software (Tabletop, 1994).
- *Teacher role:* This role contributed to actions such as intervening as a mediator in the pupil's learning, observing students' strategies, preparing pedagogical materials for a range of activities, discussing the methodology and management of IT classes, giving new statistical contents meanings, developing new professional skills, and, finally, dealing with software resources so as to use IT as a pedagogical instrument.

- *Researcher role:* The importance of taking the role of researcher regarding professional development could be assessed by collaborative work group bonding, reflection on practices, reflection on education, development of pedagogical and mathematical knowledge, development of computing skills, and finally development of self-esteem and confidence.
- *Teacher educator role:* It involved, among other aspects, interactions between both groups of teachers and handling conflicts. The beginning of the relationship was problematic and involved what Hargreaves (1998) calls "balkanization."[5] This metaphor describes the different subdivisions that can be found in an educational workplace and might cause difficulties or even jeopardize the constitution of a collaborative working group. In fact, division into two groups created a distinction of commitment between the members of each one. Teachers of the Action Group had to follow a strict schedule to become fully dedicated to the project, and had to hold meetings with researchers at the university, whereas the rest of the teachers only had to attend classes or local meetings.

Balkanization has the following characteristics: *reduced permeability*—the subgroups keep themselves completely separated, as their knowledge, beliefs, and ways of thinking become distinct; *long-lasting permanence*—few members leave the group every year; *personal identification*—a kind of "segregation" arises, allowing identification of a professional with a specific subgroup; and *political nature*—the realization that influence is a prerogative of a particular subgroup.

Since Hands-On Group teachers took part in the project on a compulsory basis, attending it during their official periods of work, it encompassed a broad cluster of participants but did not help to stimulate these teachers to become more engaged in or committed to their tasks. As a result, the group's collaboration was not spontaneous in the sense understood by Hargreaves (1998) and became "constrained."

In the school working culture, there are two different types of professional behaviour: *individualism* and *collegiality* (Hargreaves, 1998). The first type consists of private activities dissociated from those of the group, and the second type is characterized by authentic interactions between colleagues so that they can work jointly, making common decisions and planning shared actions. The dialogue allows them to exchange ideas, being involved in actions with all participants, and in general is voluntary or the result of some teachers' spontaneous initiative. It is completely different from *artificial collegiality,* which occurs under administrative demands and a compulsory joint effort between teachers, whose aim is often to implement proposals made by other professionals rather than by teachers themselves.

Balkanization and artificial collegiality became more pronounced as the project developed. Evidence is provided by analysis of the pedagogical discourse of the coordinator. According to her, one of her roles is to promote the interactions among teachers so that they can develop collective works, but interactions are hard to achieve if teachers go through their activities by themselves. They do not want to be exposed or "owe an explanation to somebody" (Violet, first interviewed). In another remark, she pointed out that, in general, teachers "do not show or even talk about their pedagogic approaches," which indicated to her the individualistic aspect of their job. Teachers are open to accepting proposals coming from meetings with the coordinator's group, but it seems that they do not easily agree to use them in the classroom. The coordinator concluded that teachers can consider these activities as useless or meaningless for pupils. Based on her observations, we can note the presence of an artificial collegiality among teachers in her school. Constrained collaboration is also recognizable, as teachers are expected to adopt pedagogical proposals that they have not initiated or a strategic line of action that they have not conceived. However good the coordinator's intentions are, or interesting the activities are, the action-curriculum remains dissociated from the planning and creation process, which could explain, in the end, teachers' considerable resistance. This was the situation faced by the Action Group members as they undertook the educational process with the Hands-On Group teachers.

At the opposite end, the mathematical discussion presented below is an example of collaborative work in which decisions are made together, providing moments of intense learning, reflection on the extension of this learning, and reflection on knowledge developed through the discussion points. Doubts are exposed and clarified, and any inconsistency between what somebody had previously thought and what he or she has obtained as feedback from the software is discussed. Figure 8.1 refers to a situation called *new issues for exploring Todos.TDB databases,* whose content is related to "proportion."

Faced with the question "In what place do we have a bigger proportion of married people: in the countryside or in the capital city?" the group suggests two distinct ways to obtain the answer. As shown in Figure 8.1,

Figure 8.1
Database TODOS.TDB fragment and graphs from Marital Status X Capital

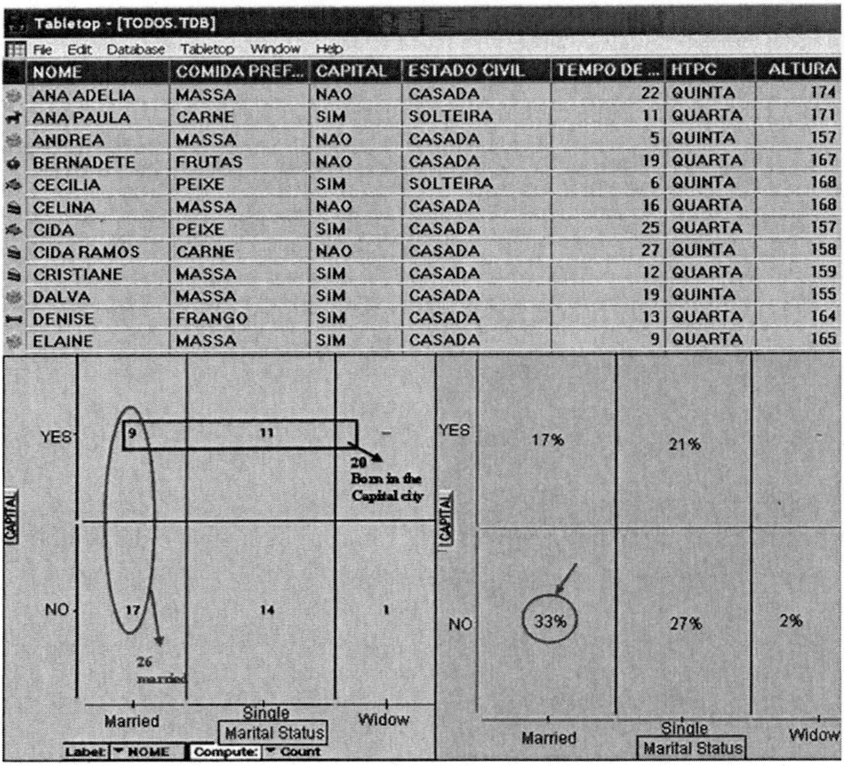

Tabletop - [TODOS.TDB]

File Edit Database Tabletop Window Help

NOME	COMIDA PREF...	CAPITAL	ESTADO CIVIL	TEMPO DE ...	HTPC	ALTURA
ANA ADELIA	MASSA	NAO	CASADA	22	QUINTA	174
ANA PAULA	CARNE	SIM	SOLTEIRA	11	QUARTA	171
ANDREA	MASSA	NAO	CASADA	5	QUINTA	157
BERNADETE	FRUTAS	NAO	CASADA	19	QUARTA	167
CECILIA	PEIXE	SIM	SOLTEIRA	6	QUINTA	168
CELINA	MASSA	NAO	CASADA	16	QUARTA	168
CIDA	PEIXE	SIM	CASADA	25	QUARTA	157
CIDA RAMOS	CARNE	NAO	CASADA	27	QUINTA	158
CRISTIANE	MASSA	SIM	CASADA	12	QUARTA	159
DALVA	MASSA	SIM	CASADA	19	QUINTA	155
DENISE	FRANGO	SIM	CASADA	13	QUARTA	164
ELAINE	MASSA	SIM	CASADA	9	QUARTA	165

somebody could conclude that there are 26 married people, 9 born in the capital and 17 in the countryside, and therefore the expected ratios are $9/26$ and $17/26$. However, the group raises a doubt, as we will see in the transcript below: given that 20 people were born in the capital, and that 9 of them are married, $9/20$ and $11/20$ might be the answers. After a discussion, the question asked to the Hands-On Group is replaced with the following one: "What is the proportion of married people born in the capital?"

Sandra Magina: Is it true that more people from the capital are married than people from the countryside?

Margaret: Within those who were born in the capital or in the interior, I realized that there were 17 married people who were born in the countryside and 9 married ones who were born in the capital, then the answer is that we have more married people who were born in the countryside. A proportion of 33% are born in the countryside, and 17 are born in the capital. What arrives at 9 to 26, here there are 26 married, 25 singles, and 1 widow. Just 2% of widows, that will be 2% of the total, then I put it in the total, but I did not put the data here, so 26 that will be 33%, 17 married.

Lulu: Because the 33% means out of 52, it is not the percentage of married ones. So the first thing you will do is to look at the numbers, sure, it is clear that comparing 26 with 9 means that you have more married people who were born in the countryside. (EN 23)

More globally, the analysis shows that the teacher educator role contributes to a "reconstruction" of the role of the educator and of education itself, related to preparing pedagogical material, acting as a mediator of the teacher's learning process, observing teachers' strategies, developing new skills in computing, and reflecting on the group's practices and theoretical background.

Conclusion

At the beginning of the in-service educational program, Action Group members gave priority to acting mainly as learners, trying to become able to use specific computing tools. Gradually, they started to take on other functions and adopted a critical and questioning posture, exposing their own didactic concerns. While they were constrained by the group, each one became engaged in preparing and developing the workshops as well as sharing decisions in a more intensive way. Afterward, the responsibility of running the workshops was exclusively that of the school's teachers.

Some characteristics of this school-university partnership contributing to the success of the Action Group's collaborative work were (1) the school professionals' voluntary participation, (2) the egalitarian relations between members of the group (i.e., its non-hierarchical characteristic), (3) the guarantee that each participant would have the chance to say something, to explain his or her point of view, (4) the establishment of a relationship based on trust, (5) the acceptance of the teachers' pedagogical practices in the group's discussions, (6) the possibility for all members of using every project's materials and documents, and (7) the presence within the school of the university's researchers throughout all stages of the project.

Based on the data analysis, the collaborative working group can be considered as a context for professional development, as indicated in Figure 8.2. The main factors that seem important in this particular process were (1) the educational program rooted in the school and planned to fit its characteristics and needs, (2) a process based on a long period of time, (3) a process aimed at the creation of learning situations and pedagogical materials, and (4) the use of computing tools associated with other educative resources throughout the process.

For the teachers' professional development, it was crucial to try to take the roles of learner, teacher, teacher educator, and researcher, because these roles allowed the members of a collaborative working group not only to increase their mathematical and statistical knowledge but also to improve computing skills, acquire autonomy and confidence, and adopt a critical posture on teaching practices as well as on the national curriculum and learning and teaching processes.

Figure 8.2

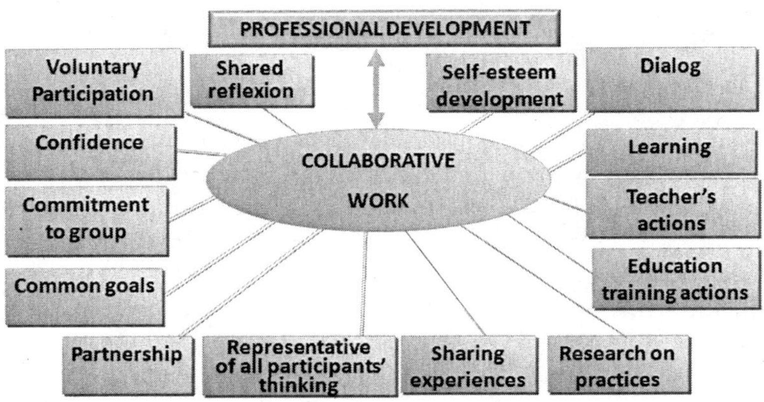

Notes

1 This chapter is related to my PhD research in the Education Post-Graduation Program, from Pontifical Catholic University of São Paulo, 2004.
2 Lulu Healy, Sandra Magina, Sandra Santos, and Nielce M. Lobo da Costa.
3 These are all fictional names.
4 Both of them, motivated by reasons independent of the project, moved away from the school after being selected in a government aptitude test.
5 The term "balkanization" was inspired by experiences in the former Yugoslavia, when the country, despite being considered as the pearl of Eastern Europe, and as a result of its breakup, evolved toward religious and ethnic conflicts, long-standing suppression of which could not prevent the outburst of a bloody war.

References

Greenwood, D., & Levin, M. (2000). Reconstructing the relationships between universities and society through action research. In N. K. Denzin & Y. S. Lincoln (Eds.), *Handbook for qualitative research,* 2nd ed. (pp. 85-106). Thousand Oaks, CA: Sage.

Hargreaves, A. (1998). *Os professores em tempo de mudança: O trabalho e a cultura dos professores na idade pós-moderna.* Lisboa: McGraw-Hill.

Mathison, S. (1988). Why triangulate? *Educational Researcher, 17*(2), 13-17.

Pérez Gómez, A. (1997). O pensamento prático do professor: A formação do professor como profissional reflexivo. In A. Novoa (Ed.), *Os professores e sua formação,* 3 edição (pp. 93-114). Lisboa: Publicações Dom Quixote Instituto de Inovação Educacional. (Nova Enciclopédia Temas de Educação, vol. 1).

Schön, D. A. (1983). *The reflective practitioner: How professionals think in action.* New York: Basic Books.

Tabletop. (1994). New computer tools for logic, information, graphing, and data analysis. *Hands On!, 17*(2), US digital document retrieved 23 February 2009 from http://www.terc.edu/handsonIssues/f94/tabletop.html.

Chapter 9

Formative Processes: Sharing Learning Experiences in Geometry

Adair Mendes Nacarato and

Regina Célia Grando

Introduction

This chapter is related to research developed from 2003 to 2007 in a collaborative group constituted by graduate students (pre-service mathematics teachers) and postgraduate students, in-service elementary school teachers, and university professors. The group's object of study was geometry and its teaching; it was this focus that motivated the name of the group: Collaborative Group in Geometry (Grucogeo). The group met weekly during two-hour sessions in the university. It encouraged, in each semester, voluntary participation, with possible entries and exits of participants, having on average fifteen participants. In this space, five researches were developed: one coordinated by the authors of this chapter (with external financing from MCT/CNPq[1]), two scientific initiation researches, and two master's degrees in education researches.

The look at geometry and its teaching was diversified: exploration of investigative tasks with and without the use of software; experimentation with the same task with different media (in Grucogeo or with elementary school pupils); exploration of the potentialities and limits of each medium (folding, constructions with ruler and compass, use of manipulable materials and computer environments); and discussion and analysis of the argumentation, validation, and proof processes in geometry.

The proposal of the work group centred on the elaboration, development, evaluation, and analysis of activities for elementary education. The group focused on some activities initiated by participants (in-service teachers and university educators), some of them being developed by the university educators. These activities were re-elaborated in a way that could be used with pupils in the school classrooms of the in-service teachers. Sometimes pupils worked in the institutional space of the university.

For the classroom activities, participants were divided into subgroups in such a way that, in each subgroup, there was at least one elementary teacher to develop the lesson with pupils along with graduate students to register the process and management of the activity. The data (productions of the pupils, daily field diaries of the graduate students/researchers, and narratives of the teachers) were discussed, analyzed, synthesized, and systematized in Grucogeo.

Underlying Framework

This work proposal was based on the use of formative strategies (Alarcão, 2003, p. 58), whose "objective [is] to make the teachers more competent in the analysis of the questions that surround their routine, allowing them to act, helping them not only to solve immediate problems but also to situate them in a broader horizon that gives a perspective to their action and to the function of school in the society where we live."

Among the strategies suggested by Alarcão (2003, p. 51), in Grucogeo we used "the questioning of other educative actors," "the confrontation of opinions and approaches," "the discussion group or study circles," and "the pedagogical questions." There was in the group a culture of questioning and problematization of the questions relative to geometry and its teaching.

We also adopted the strategy of production of oral and written narratives about daily practices of the teachers (Alarcão, 2003). Cunha (1997, p. 186), referring to the narratives as formative processes, affirms that, "at the same time that the subject organizes the ideas for the—written or oral—narrative, he or she reconstructs his or her experience in a reflective way and, therefore, makes an auto-analysis that creates new bases of understanding for his or her own practice. The narrative provokes changes in the way people understand themselves and others."

We also started from the supposition that the heterogeneity of the group—composed of in-service teachers and graduate students (pre-service teachers)—and the formative strategies would contribute to the construction of a repertoire of knowledge and professional development of teachers.

Although future teachers have left school recently, in the initial training process, they need to transform themselves from having a student's posture to having a teacher's posture. This transformation implies a change in the way of looking at school and classroom culture. From this perspective, in-service teachers assume the role of educators and co-constructors of future teachers' professional identity, in a process of mutual involvement, using "guided participation" (Rogoff, 1998). Rogoff considers that this participation can be "tacit or explicit, face-to-face or distant, involved in joint efforts with people. ... It includes deliberate attempts to teach, and incidental commentaries or actions that are heard by chance or seen, as well as the involvement with materials and particular experiences that are available, indicating the direction in which people are encouraged or discouraged to go" (1998, pp. 129-130).

The teaching activity is culturally organized; the classroom is a space that the in-service teacher knows and dominates, with its rhythms, rites, and symbols. Therefore, he or she can contribute to developing knowledge about this space for future teachers. In fact, the heterogeneity of the group permits different actors to assume distinct roles in a process of mutual help. In-service teachers bring their practices to the discussion so that they can be reflected and questioned, reaching new meanings and assuming, together with the university educators, a role of formation for the future teachers who, at the beginning, participate in Grucogeo in search of specific geometrical knowledge. The sharing of ideas by these graduate students (pre-service teachers) and the in-service teachers allows the appropriation of professional teaching knowledge by future teachers. On the other hand, graduate students, immersed in the academic culture, during the confrontation of opinions and approaches with teachers, contribute to different ways of thinking about geometrical processes (mainly in relation to proof and validation, from which teachers are distant) as well as using computer media. The university researchers/educators contribute to the group, at the same time as they appropriate the practices of elementary school teachers, to bring a researcher's point of view, with his or her theoretical and methodological experience and knowledge, explaining problems and asking pedagogical questions.

This existing movement in Grucogeo allows us to recognize the collaborative dimension of the group, which enriches the learning and professional development of teachers. Having this in mind, in this chapter we will analyze the indications of learning in, from, and about teaching (Cochran-Smith and Lytle, 1999) and the professional and teaching development of participants. We will present one of the experiences of the group in which we underline the moments when participants acted as apprentices, as teachers, and as researchers. Different data were recorded for the research: audio and video recordings of the group meetings; video recordings of the pupils' solving activities during sessions at the university; written narratives of the teachers; and written registers of the group meetings.

The Triangular Inequality Activity: Sharing Learning in the Group

Triangular inequality appeared when the university researchers/educators suggested to the teachers, starting from the content that they would teach in their classrooms, to think of some situation that could be worked with the support of different media.

Teacher Paulo pointed out his desire to work with the property of triangular inequality. At that moment, the group, in the position of apprentice, searched for an explanation of this property; mainly, the graduate students did not know it (it is important to underline that these students were products of a Brazilian school system that, for many years, had abandoned teaching geometry). However, in this initial discussion, mathematical and pedagogical knowledge were connected, because for the graduate students it was a moment of learning the geometrical property and for the in-service teachers there was concern over how to treat in a pedagogical way such content.

Paulo suggested that the first medium could be a game with three dice and sticks with measures going from one unit to six units of length. For each measure, there would be three equal sticks. The pupil would throw the three dice and grab the sticks corresponding to the values designated by the dice. For example, if the dice indicated three, four, and six points, the pupil would grab these three lengths of sticks. The objective is to verify if the three sticks allow one to form the representation of a triangle. The pupils would register their own conclusions in a table format, or in other formats, searching for a generalization of the property.

The group thought that the activity was pertinent and gave suggestions related to organization of the material. It was also consensual in the group that this would be carried out with different media (the game, the use of Cabrí Géomètre software, and constructions with ruler and compass). As previously underlined, the heterogeneity of the group allowed the valorization of different ideas. For example, postgraduate student Jorge, using his experience as a mathematics and computer sciences teacher, presented suggestions on how this activity could be developed with the software.

After planning the activity for the fifth and sixth grades (eleven and twelve years old) of the public school in which Paulo worked, it was agreed that it would be followed by the Scientific Initiation Scholarship graduate student Thiago, also a participant in the group. They would act as teachers in the classroom.

After working with the game, together with Thiago, Paulo brought to Grucogeo the pupils' productions so that they could be discussed. In the oral narrative of these lessons, he pointed out some questions that prompted reflections in the group: on the non-familiarity of the pupils with group work and resolution of open activity; on the difficulty of understanding the problem's proposal, demanding constant intervention by teachers; on the difficulty for pupils in writing and explaining their strategies; and mainly on the conceptual gap observed in relation to the concept of the triangle.

Because of the conceptual gap among the pupils, Paulo asked them to answer "What is a triangle?" The answers generated discussion in Grucogeo, mainly from a question posed by Paulo: "How can I intervene to supply this conceptual gap; in other words, how can I deal with pupils' conceptions?" The discussions centred on elements such as the inadequate use of mathematical language; the prototypical geometrical object triangle; and the necessary and sufficient conditions for the existence of a triangle. These elements of discussion emerged from teaching geometry, which, when it is present in the classroom, is based on nomenclatures from meanings and stereotyped geometric figures. These practices prompt pupils to conceive of a triangle as a figure with two equal sides and a different "base" (isosceles triangle) or as a figure with three equal sides (equilateral triangle), always in the same position, calling the side parallel to the line in the notebook the base. Pupils could not advance in the discussions about the property of triangular inequality because they did not admit the existence of the scalene triangle.

This discussion allowed reflections of two different natures in the group: epistemological questions related to the concept of the triangle and to geometrical thought, and pedagogical questions related to the challenges and ways of facing the difficulties and conceptual gaps presented by pupils. Such gaps were put in light in the use of an open activity and its transcript.

The group, on this occasion, presented two suggestions to continue the activity: (1) returning to the pupils some of the answers registered about triangular inequality, and (2) continuing the activity with use of the Cabrí software.

In this second stage, Paulo returned to the classroom and presented to the pupils three productions about triangular inequality, and he proposed that they analyze these solutions and choose one as a generalization of the property; for each case, he suggested that the pupils present examples and counterexamples. This activity was complemented with use of the Cabrí software, starting from the suggestions given by Jorge: draw three segments and the triangle constructed with these segments; with the use of the mouse, the pupils had to move the vertices of the triangle, getting new measures for the segments, and complete a table with the obtained data, searching the regularity and, consequently, the generalization.

In the discussion in Grucogeo of the second activity, Paulo underlined that the group work and the arguments developed by pupils, confronted with the presented definitions, allowed them to approach the property of triangular inequality and to extend their conceptions of the triangle. According to Paulo, even with the use of Cabrí, the argumentative answers of the pupils were still incomplete. In the concrete material (the game), they were close to the inequality relation, but these results were not transferred to the computer environment. This means that the pupils were not able to link the two media. On the other hand, use of the Cabrí software allowed the use of numbers that were not whole (it was a limit of the game), opening spaces for the manipulation of measures in the set of rational numbers as well as contributing to the generalization of the property. The group pointed out that the media, in a certain way, complemented each other: the limits of one medium were surpassed by use of the other medium, strengthening the importance of pedagogically working with different media in the school environment. It was then suggested to Paulo that the computer medium also has limits (e.g., size of the screen) and that consequently another medium could also be explored: the ruler and the compass. Moreover, for mathematics teachers, there remained the necessity of searching for a formal validation—demonstration of the validity of triangular inequality—and this stage was not taken into account in the classroom due to the grade levels of the pupils. However, in Grucogeo, the participants, again as apprentices, studied this demonstration.

Some elements of this activity were objects of reflection for the group.

1. The fundamental character of the elaboration of a written production to discuss the procedures of mathematical validation: the capacity to elaborate well-based arguments is essential for the mathematical proof process.

2. The investigative position of the teacher, searching for ways of questioning pupils without supplying ready models: this demands time of the teacher, and its insertion into the curriculum implies a review of how the educational program is organized, with little space for the teacher who would like to carry out such activities in class.
3. Even if the pupils did not manage to generalize and find the law to construct any triangle, the task was very rich, leading to new questions: on the concept of the triangle; on the concept of shape (what is shape in geometry?); on the writing process (the pupil is not motivated to write because he or she does not see the pertinence of doing so and has no reader for the text); perhaps this can open debate on a new mathematics classroom culture. As Paulo underlined, most of the time he does not know what to do with pupils' writings.
4. The pedagogical intervention was also a focus of discussion: in which moments do we as teachers intervene, and in which moments do we leave pupils alone to search in their own ways?

Indications of Teachers' Learning about, in, and from Practice

Analysis of the experiences, when explained to the group, allowed moments of reflection and discussion about the teaching of geometry and about the mathematics classroom culture. Different elements were pointed out.

1. The importance of giving voice to pupils: these are moments when we can identify concepts that were already or not appropriated by pupils.
2. The importance of pupils' productions and what to do with them: writings can be returned to pupils for analysis of the underlying arguments and can be references for the teacher in evaluating how pupils appropriate the concepts being worked.
3. Alternative ways of favouring conceptual elaboration in the classroom: we believe that involving us in the task, searching for hypotheses, arguing for validations, discussing activities with colleagues, and searching for consensual syntheses in the classroom constituted authentic mathematical activities.

In the written narrative about his experience, produced for presentation in the Sixteenth Congress of Reading, June 2007, Paulo noted this:

> The written productions were fundamental to identify the existing gaps in the conceptual elaboration of the triangle by pupils and to be able to make adequate interventions. This process was enriched by moments of negotiation of meanings, where the pupils argued in favour or against a determined hypothesis. In this dynamic, the mathematical truths are validated by the pupils and not imposed by the teacher. ... The participation in Grucogeo and the possibility to share my doubts, uncertainties related to a new approach, as well as the systematization of the experience in the present text were fundamental to my professional development.

Another question present in the studies of Grucogeo was related to the importance of using different media in the teaching of geometry. Evidence of appropriation of these discussions, knowledge produced on this question in the group, as well as adoption of the same media in the practices of future teachers are found in the narrative produced by graduate students Valéria and Renata.

> When one works with different didactic resources in the classroom—it was the objective of our project of intervention—the pupil constructs and produces knowledge about the different representations approached by the different media. It can be said that these passages of one representation to another allow the pupil to develop abilities and to start to appropriate a concept. ... We recognize that our "bravery" in working with computer environments in the classrooms was a consequence of the shared experience coming from Grucogeo.

From the perspective of Bolzan (2002, p. 148), the group allowed the weaving of a set of relations in which "a voice joined the others, and, in the course of the interactions, meanings and ideas about pedagogical knowledge were shared." The group constructed a pedagogical knowledge on geometry through shared and systematic work. The statement of teacher Mirian in an online interview, December 2005, supports such a construction:

> The geometry workshops [Grucogeo] allow me to explore the biggest number of properties, regularities that exist in geometry. How many things that I saw in a definitive way and that other people saw in another

> way; this makes me extend my knowledge, and I can deal with these questions in the classroom, where there are forty heads reasoning differently from the others.

In this construction, everyone appropriated the meanings produced by the group, transformed them, and developed them professionally. As Rogoff (1998, p. 136) points out, "a person who participates in events modifies them in such a way that it becomes different in subsequent events." Thus, the development is seen as dynamic and the appropriation as "a transformation process" (p. 134). The graduate students (pre-service teachers) appropriated strategies built by the group that they would reinvest in classrooms (they developed in this participation knowledge about and in practice); the in-service teachers built a system of their experiences in discussing them with colleagues working in other institutional spaces, becoming researchers on their own practice (they developed knowledge in and from practice); and the university researchers/educators constructed new theoretical and methodological bases for research in teaching geometry (knowledge constructed in and from practice).

Note

1 MCT/CNPq: Ministério da Ciência e Tecnologia (MCT)/Conselho Nacional de Desenvolvimento Científico e Tecnológico (CNPq).

References

Alarcão, I. (2003). *Professores reflexivos em uma escola reflexiva.* São Paulo: Cortez.

Bolzan, D. (2002). *Formação de professores: Compartilhando e reconstruindo saberes.* Porto Alegre: Mediação.

Cochran-Smith, M., & Lytle, S. L. (1999). Relationships of knowledge and practice: Teacher learning in communities. *Review of Research in Education, 24,* 249–305.

Cunha, M. I. (1997). Conta-me agora! As narrativas como alternativas pedagógicas na pesquisa e no ensino. *Revista da Faculdade de Educação, 1-2,* 185-195.

Rogoff, B. (1998). Observando a atividade sociocultural em três planos: Apropriação participatória, participação guiada e aprendizado. In J. V. Wertsh, P. Del Rio, & A. Alvarez (Eds.), *Estudos socioculturais da mente* (pp. 123-142). Porto Alegre: ArtMed.

Chapter 10

Collaborative Work and the Professional Development of Mathematics Teachers: Analysis of a Brazilian Experience

Ana Cristina Ferreira and Maria Ângela Miorim

Introduction

In Brazil as in other countries, efforts have been made toward the education of mathematics teachers (pre-service and in-service). In past decades, different initiatives and proposals were presented in the context of the reform of mathematics education. However, many of these experiences were not successful. From our point of view, the perspective adopted has not always been the most appropriate. Often these projects try to implement courses developed in other places (by governments or universities) with the belief that they are the best solutions to the problems and difficulties encountered by mathematics teachers. In these approaches, the teacher is not perceived as a partner, as a professional with knowledge, valorized experiences, and ideas. We believe mathematics teachers need to be listened to and invited to choose the areas and topics they want and need to study.

With this aim in mind, we propose an alternative for the education of practising teachers based on the constitution of a researcher and mathematics teacher group, in which participation is voluntary and decisions are made collectively. We think that, as teachers improve their knowledge and face various possibilities in developing their practice, they become more aware of the conditions for their development. University researchers and school teachers can be involved in exchanges of knowledge and practice, contributing to the construction of both new knowledge and new practice.

Conceptual Framework

Professional development is here understood as a non-linear process, variable in time, that occurs in relation to individuals' teaching experiences and mathematical learning. This process involves pre-service education and continuous formation as well as experience both as a student and as a teacher. It is marked by personal, motivational, social, cognitive, and affective factors (Ponte, 1998). The personal characteristics of the individual—her stage of life, her personality, her motivation to change, the support she receives or the pressure she faces in society, and her own cognition and representation (beliefs, values, objectives)—play an important role in this process. We can say that professional development involves a dialectic between a teacher's beliefs, types of knowledge, and previous experiences and a new type of knowledge that affects teaching practice (Llinares, 2000).

Our proposal aims to create an environment that acknowledges teachers' knowledge and practices, starts from their own objectives and expectations for mathematics teaching and learning, and, from the new information discussed, allows them to decide to restructure their practice. This is a teacher's first step toward becoming an agent of his own learning. The core of this inquiry is to build a working practice supported by both (elementary and high) school and university cultures so as to generate new knowledge for teachers and to contribute to their professional development in this process of improving mathematics teaching and learning. The collaborative group seems to be an appropriate context to do so.

A collaborative group is here understood as a group in which participation is voluntary, in which all individuals involved search for professional growth, share trust and respect, support the group work, engage in a common purpose, all the while creating and sharing meanings about what they are doing, about their lives and professional practices. In such a context, participants feel comfortable expressing their ideas freely, listening to criticisms, and changing points of view, and the activities do not follow only a determined orientation. Participants can have different levels of involvement, different interests and viewpoints, thus contributing to a great variety of ideas.

Methodology

A careful study was designed that included a detailed plan to take account of the results through presentation of the data and description of the processes.

Participants

Besides the researchers (the authors of this chapter), four elementary and high school mathematics teachers working in public schools in the city of Campinas were engaged in this study: Iva, Maria, Fernanda, and Andrea.

The youngest teacher of the group, Andrea, had graduated from UNICAMP two years earlier and taught mathematics and physics at both elementary and high school levels in both public and private schools. The isolation she felt in the schools where she was working, and her willingness to improve her knowledge of mathematics, pushed her to join the group.

Fernanda had a degree in mathematics education from Unicamp and had been teaching for four years in both public and private schools at various levels. She frequently attended courses and workshops in search of opportunities to learn new methodological strategies and alternatives. Her participation in the group reflected her willingness to exchange experiences and to create collectively new alternatives.

Iva had been working as a teacher in public schools for twenty years. In 2001, she taught in different schools, moving from one to another every day to teach morning, afternoon, and evening classes. Being part of a study group and creating activities were not new to her. She had been thinking about such opportunities for a while and had discussed them with Maria.

Maria was also a public school mathematics teacher, having worked for twenty-five years. She always taught mathematics in public schools. Planning to retire soon, she still felt the necessity and desire to remain up to date and to discuss the difficulties and challenges of classroom instruction.

Our decision to begin a study of this nature, in collaboration with public school teachers, stemmed not only from their own backgrounds as students and public school teachers but also from the importance they attached to a public school of quality for the country and to a project aimed at professional development of mathematics teachers.

Data Collection

The group had thirty meetings between January and December 2001 that lasted three and a half hours each. The dynamics of the meetings included study of a mathematical content knowledge previously selected by the group and exchange of experiences and elaboration of methodological alternatives related to the development of this mathematical content knowledge in the classroom, supported by the report of the teachers' and researchers' experiences. The data were collected from various tools: transcripts of the meetings, field journals, narratives produced by the teachers, and written evaluations.

Data Analysis

We used case study as a methodological strategy for the analysis. The case studies involved in this research were developed using two distinct but complementary units of analysis: teacher and group. We proposed to understand, on the one hand, the process experienced by each teacher during her participation in the group work and, on the other, the process of emergence and development of the group as a whole. We then created four case studies involving the teachers and one involving the group. In this chapter, we explore the case study involving the group only.

In the elaboration of that case study, we used triangulation of both data sources and researchers. In the former, we compared and contrasted the data collected with the various tools we had; in the latter, we did independent readings of the data collected so that we could compare and contrast them, discuss our observations, and produce an in-depth analysis of them. This required repeated consultations of the data collected and elaboration of various versions of the case. In this process, we were supported by all of the information collected during the year as well as by our own notes. We also analyzed our own participation as researchers and group members and our influence on the group. At this point, we tried in the literature to develop a better understanding of the elements that we thought could facilitate or limit the process of group creation, having in mind that we would analyze them. We followed our own investigative question and whatever arose from it. Other elements of interest in the data were equally considered. With this approach, we established a form of dialogue with the literature not only to shed light on the similarities and differences between our findings and those of other researchers but also to establish a theoretical reflection on the process.

The Collaborative Group: Processes and Results

Our group was not a collaborative group at the beginning. Collaboration increased gradually. In the beginning, though we encouraged the teachers to present their ideas and concerns, they expected us to prepare the materials and organize the meetings. Even if our objectives were collectively decided, we wanted to establish a collaborative relationship in which autonomy and the power to decide are minimized if there is not full participation. That the teachers were not responsible for organization of the meetings limited in a certain manner their participation, although they showed commitment to and engagement in all the activities. However, the preparation and organization of materials by researchers were natural and expected. Teachers needed more time to get to know each other and develop some level of trust and understanding of the roles in the project.

In our attempt to detail the analysis of the work developed by the group, from cooperation to collaboration, we observed distinct movements during which the dynamics and the form of participation of the group members varied. We call those variations movements because they were not static, isolated, or only connected to specific moments of the process.

The first movement began when the teachers were contacted and when the first meeting was held, in January 2001. This movement covered the group's first meetings. The teachers voluntarily joined the group and stated their willingness to engage in group work. During this period, we observed some empathy between participants as their conviviality became more natural every day. Their participation in the group was marked by increased engagement and attention during the meetings. There was only faint participation in the search for materials, in suggestions for readings, and in the presentation of pedagogical practices. We, the researchers, on the other hand, were deeply involved in the careful preparation of the activities and of the various moments of the meetings. Our participation was marked by a friendly and respectful welcome and by the explicit support offered to the teachers. We encouraged the group members to express their viewpoints, to share their experiences, to reveal their anxieties: that is, we encouraged and valued the presence and engagement of the teachers in the group.

This movement concentrated on the first studies on fractions, as collectively defined. During the activities, we tried to create an environment in which no one was an expert, in which making mistakes, not knowing, and having doubts were allowed. The belief within the group was that what counts is to be aware of one's own limits and to try to overcome them, either by asking for someone's support or by finding support in a text every time there is a need to develop knowledge.

> I'll tell you the truth. Do you know how I managed to get an answer? Making calculations on a sheet of paper and introducing them. ... What happened was that I found the result of the division, but I could not understand how I should register the answer, then because of ... what I did, then I took it and did it again, I did it in the. ... But it is not. ... You think so, but I was not able to do it with the card. (Fernanda)

> I couldn't either. You know that I tried, but in the division and in the multiplication I couldn't do anything. (Iva)

How we organized the first meetings was crucial for the participation of the group members throughout the process. When the teachers felt they were accepted, respected, and valued for their experiences and ideas, they became more confident in themselves, in the group, and more ready to share their ideas and face criticisms.

However, those elements were not the only ones, nor the most important ones, in allowing the process of change to occur. The individual characteristics of the teachers, how they interacted with each other and with us, and the challenges and activities proposed greatly influenced the whole process. However, empathy and affinity cannot be created or imposed. In this group, we were lucky enough to join people whose individual characteristics—sincerity, honesty, commitment, affection—were extremely favourable to collaborative work. Therefore, the links between members of the group became stronger, and the group slowly became a collaborative environment.

From the beginning, the dynamics of the group were supported by activities proposed by the group and guided by us (taking account of the analysis of the group process). Moreover, it is possible to perceive that the dynamics of the group—its practices, routines, activities, and readings—were also influenced by theoretical elements. During the meetings, time was dedicated to the reading and study of papers as well as to discussion related to proposals and materials coming from a review of the literature.

A movement began during the first meeting and gradually became more evident. One specific activity, which occurred between April and June, greatly influenced evolution of the collaborative work group. The activity was centred on the investigation of students' knowledge of fractions.

As we could see, that the group decided to investigate and accept all of the difficulties and consequences of the tasks they proposed drastically changed the dynamics of the meetings, contributing to more active participation of the teachers. This was due to the characteristics of the activities, which demanded discussion and step-by-step restructuring, from the objectives to the activities themselves, from the classrooms involved in the project to the analysis.

The process involved the teachers in discussions and reflections about their own practices and their consequences for students' learning and about the elaboration of objectives and evaluations. Questions related to the use of adequate language, to the explicitness of the information presented, and to the expectations related to students did not seem familiar to them and therefore demanded great effort.

Teachers learned a lot about how to investigate students' conceptions of fractions as they attended different school levels or how to analyze students' mistakes and learn from them. However, the essence of this experience was the opportunity given to teachers to be able to investigate, to share tasks, to work individually, in pairs, and in the group, or to produce work representative of (and important to) the group. This type of work made all the difference. After this process, the group was not satisfied to merely wait for interesting ideas to come along and develop activities. The group members were ready to take control of the meetings and of their own learning.

During the period of investigation, we observed that the group's position was consolidated and evolved toward collaboration over time. In the beginning, the teachers seemed to expect some type of guidance, some orientation about what to do. Slowly, they started to play their roles as investigators and as generators of knowledge. The course of the research depended entirely of the decisions and work of the group. At the end of the research process, the elements that characterized the first period had disappeared: dependency and poor mobilization in the process of making decisions and choices, of criticizing and pondering. The group began another type of relationship, that of collaboration. This experience proved to be an important learning opportunity that led all participants to the third movement: consolidation of the collaborative group.

At the end of the investigation, the teachers had already adopted another attitude. Participants were more confident and active in regard to their choices and decisions. We still contributed materials, but the teachers now brought their own materials, which included interesting selections from texts and activities available to them, which they carefully organized to study and use. At the end of the first six months, we evaluated the work of the group. The teachers' discourse on the meetings helped us to see the contribution of the group.

> In what concerns the dynamics of our meetings, I can only praise them. ... (c) I feel good when I feel that I can complete an activity ... (e) and find that this homework is really good. ... It is like a link, and I feel motivated for the next meeting. ... (f) I love to receive copies of the articles. ... I feel important when I have a text in my hands, which will help me improve my work. (Andrea)

> The work is living up to my expectations due to the harmony and synergy ... among us. ... One great point was the presence of Ana Cristina in the classroom, because it allowed me to resolve my doubts and to have some auto-criticism about my way of being. (Fernanda)

> I found that it is necessary to search continuously about mathematics content knowledge and not simply follow the lessons in a book. ... I feel that everybody is anxious to grow and that it was worth engaging in the group, because I have learned a lot from everybody. (Iva)

> The meetings have been encouraging me to review my classes and find what to change in my way of teaching. As I am not able to do it immediately, these ideas will remain in my mind, so that during the second semester I'll apply these ideas to my classes. (Maria)

In the course of the second semester, group consolidation intensified, and at the end of the year many activities were held simultaneously. The group had reached a stable level of autonomy, and all activities were now feasible and efficient. Some of the elements, more or less evident in each one of the previous moments, became essential to the changes observed in the group. These elements were the work done with content knowledge, its didactic aspects, the experience with activities, the observation of the teachers' classrooms, the introduction of innovative activities in the classroom, and the production of episodes. They contributed to the development of professional knowledge about mathematical content and didactic aspects of teaching and learning this content. We observed that the teachers referred to them during the year in different ways. Initially, the tasks were performed with interest and curiosity, but the teachers were not so critical of their own selection and organization of the tasks. After some time, they started to analyze the activities in relation to the realities of their classrooms (students' characteristics, resources available, etc.). Finally, they were critical, analyzing each activity, detecting problems, observing inconsistencies, seeing potential, allowing them to implement, in their own spaces, the activities that seemed suitable to the needs of their classrooms.

> Then we were able to see and discuss all this. I liked the class, its involvement. ... And I realized that they were involved also. They were, you know, all asking and participating. Not a hundred percent, as I told you, no. But like that, in a regular class, if it was not like that, it was. ...

> And students who had never participated before became engaged. And finally they said, "Teacher, what a class you taught. I like it this way." ... And they asked for other classrooms like that. (Iva)

Those moments were important for the research. All the members of the group learned from the reports of the teachers who had done the experiment. Our knowledge of the students increased. The discussion opportunities encouraged the teachers who were more resistant to try to innovate too.

> Then I got there and said, "Look, today we are going to work on a different subject. ... You're going to read everything in detail." ... "Hey, what's a scale?" ... "What should I do? How does this thing work? How do I measure that?" ... I had to move from desk to desk to help them do the measurement. ... Even so, they asked too much, too much. ... So I found it complicated. ... Nobody said, "Oh, that's cool!" Nobody said so, ... and I had to reprimand some four pairs of students there. ... It was terrible, then I had to complain. ... Now that's what followed: I found it kind of positive to have them interested in showing the value here around the contour, measuring, showing that they were learning. So I found it positive. (Maria)

Analysis of the activities helped their adaptation and implementation in the classroom, and the production of classroom records was extremely valuable in looking back at the activity in the group through reflection and analysis by the teachers of their own practices and those of others. Producing the records allowed the teachers not only to review the classroom activities but also to realize their intentions while reviewing their actions in regard to the results obtained. Moreover, participants' metacognitive reflection was developed when the group discussed texts. Participants had, besides their own viewpoints, those of somebody else for whom they felt affection, trust, and respect. All this contributed to a more conscious view of themselves as teachers, of their pedagogical practice, and of different possibilities and alternatives for that practice.

Another element, classroom observation, also played an important role. Taking account of others' views, ideas, and partnership increased the possibilities offered by the group. The group was there not only studying and discussing the reality of schools together but also met at the teachers' workplaces, observing and experiencing the limits, difficulties, and potentialities of the activities.

However, the process of learning and changing was not equally fulfilled by all members of the group. They accomplished distinct tasks according to experience, life development, personality, professional background, and individual characteristics. Whereas for Andrea the group proved to be an opportunity to increase self-esteem and self-confidence, for Maria it meant a moment to review her professional history in light of changes in the educational system and in regard to behaviour of the students, which opened up the possibility to modify her beliefs concerning the roles of content knowledge of teachers and students. In other words, the experience of participating in the group had specific and personal meanings for the teachers. This richness was clearly seen in the case studies.

> I began to question the planning and distribution of the subject matter, whereas before I followed to the letter what each school dictated. I did not have any longer a resistance to reading articles about mathematics education that I felt were too difficult to understand. I write the objectives for each class. I think of different ways to teach the same subject matter. I had such attitudes only superficially during my undergraduate studies, but today I feel comfortable in applying them. (Andrea)

> I think that everything needs to be well thought out before going into the classroom. ... This is reinforcing each time. I believe that today I do this, sit down and prepare. For example, the weekend I prepare, think, sketch out my lessons, like you always say. (Iva)

> I was always preoccupied with the way students thought, but the survey helped me to see that we should, at least from time to time, listen to and understand the way the students see things. I realized that in some activities students who seemed distant many times needed only a little motivation and attention. (Fernanda)

The above transcripts indicate metacognitive development. Understanding the importance of students' knowledge and reflecting on respect for their needs helped the teacher to make better-informed decisions as an educator. Fernanda realized that listening to a student is much more than paying attention to what he says and to his reality; it also involves an understanding of his mathematical thinking, his way of reasoning, his knowledge, et cetera. At the end, she evaluated her actions having in view not only goals for the students but also for herself. These are significant steps. Iva recognized and emphasized the importance of planning her lessons. This was seen as being essential to support students' learning. Another aspect developed was her curriculum knowledge and her role in relation to that curriculum. Andrea discovered that the program and the curriculum were not (or should not be) rigid and close; they are not a linear and progressive sequence of "prerequisites." She began to understand that the teacher has a degree of autonomy, that she is the authority in the classroom, and that she has the power to decide the order and the manner of developing the subject matter.

During the first semester of 2002, anticipating her retirement, Maria showed that she was sensitive to her identification with the group's work the previous year. We learned through telephone conversations with her and from accounts from Iva (whom Maria encountered each week) about her attempts at innovation as well as her enthusiasm in her grade 6 classes (things that had not happened in previous years). According to Iva, "Maria went up and down the stairs of the school with sacks full of teaching materials and was very animated with her sixth-grade classes!" (personal communication). It is interesting to highlight some elements of this apparently casual comment. First, as we said, Maria did not want to teach classes of students in the early grades because she had questions regarding the significance of many concepts. She became involved and interested in working with fractions in the group; however, besides the survey activity, she did not have opportunities to apply this knowledge.

The collaborative group became a space in which each teacher brought expectations, experiences, and epistemological points of view and searched to extract answers. It formed a space in which to deepen content knowledge, to develop and analyze a project, a proposal, a lesson, as well as to listen and be heard, to grow, to learn, to move.

Conclusion

Research is not always easy when the researchers themselves are involved in the group through direct participation. However, if the teacher is considered as a partner in professional development, somebody who possesses practical pedagogical knowledge, then such a strategy becomes necessary. It becomes part of the research and can give input into the process of improving education and learning.

From this perspective, collaborative work is an interesting and fruitful avenue. In our group, teachers and researchers worked together, each one coming from her own particular reality, history, and development as a professional, sharing knowledge, and constructing new ways of conceiving and developing education and learning. The group confirmed that those processes demand more time than we initially thought they would take. Rooted in a complex context, the culture of the classroom and, more specifically, the pedagogical practice of each teacher do not change so quickly. There is an initial moment when teachers learn for themselves and about themselves. They increase their knowledge of mathematical content and its didactics, of students, of themselves. Through this process, a teacher's awareness of her classroom, aims, and intentions is also developed. However, reflection on pedagogical practice does not quickly become a new way of working, a new way of viewing interactions between teacher and students, a new way of viewing the curriculum and mathematics itself.

The success of this group was partially due to the fact that participants encountered conditions favourable to this type of work: adequacy of the group size, regularity of the meetings, which occurred on dates collectively established, and implicit conditions (e.g., the accomplishment of tasks). We implemented a collective action plan: that is, the meetings were organized to reach goals decided by the group. The group constantly encouraged the presentation of individual and collective records and reflections. All members had the same status (participants in this project were entitled to share their opinions about everything and to express their ideas freely while others listened attentively). Attention was focused on the teaching and learning of mathematics, and periodical evaluations were applied (Murphy & Lick, 1998). Moreover, in contrast to research that defends teacher group formation (Antúnez, 1999), in this group the teacher-researcher and school-university partnership was constructive and positive. Evidence suggests that, depending on the characteristics of members of the group and on their collective intentions, this partnership can bring great benefits to all.

Another positive outcome of this work was transformation of the teacher-researcher relationship into that of a true partnership. Mutual learning became possible through the model outlined in this chapter. It gave value to the social interactions in the work of both teacher and university researcher. If the goal is improvement in teaching and learning mathematics in Brazilian public schools, then it is fundamental that theory and practice are analyzed on the ground in schools through collaborative actions as demonstrated by this project.

We have pointed out the evolution in thought and practice among the teachers, but we are aware of how much we, the researchers, learned in our interactions with them. We also learned a lot by experiencing the classroom realities of these teachers, increasing our knowledge and understanding of the school setting in terms of teaching and learning mathematics. The teachers were important for our own professional development as researchers. This aspect requires more analysis in future research. We would like to know more about the contribution of collaborative work to the professional development of the university researcher belonging to such a group.

References

Antúnez, S. (1999). El trabajo en equipo de los profesores y profesoras: Factor de calidad, necesidad, y problema. El papel de los directivos escolares. *Educar, 24,* 89-110.

Llinares, S. (2000). Formación de profesores y educación matemática, conferência pronunciada no ato solene de inauguração do curso 2000-2001 na Facultad de Ciencias de la Educación de Melilla, Espanha, 6 de octubre.

Murphy, C. U., & Lick, D. W. (1998). *Whole faculty study groups: A powerful way to change schools and enhance learning.* Thousand Oaks, CA: Corwin Press.

Ponte, J. P. (1998). Da formação ao desenvolvimento profissional. *Actas do Profmat 98* (pp. 27-44). Lisboa: Associação dos Professores de Matemática.

Chapter 11

Assessing a Research/Professional Development Model in Patterning and Algebra

Ruth Beatty and Catherine D. Bruce

Introduction

Over the past fifty years, researchers have studied the importance of teacher professional development (PD). What teachers know and are able to do is a key factor influencing student learning (Fullan, Hill, & Crevola, 2006). However, there is a lack of empirical research on the effects of PD on teacher learning, particularly with respect to developing classroom knowledge and improving practice (Hill, 2004), which are crucial indicators of success (Fishman, Marx, Best, & Tal, 2003). Our goal was to develop a Research/PD Model that embedded professional development within a longitudinal research program for two reasons: (1) to continue to evaluate an innovative approach to teaching patterning and algebra in elementary grades; (2) to assess the effects of including teachers as collaborative partners in their development as mathematics educators. To this end, we examined teacher learning and its links to teacher participation (with a critical focus on teacher enactment), teacher knowledge, and classroom practice.

Background to Patterning

Patterning activities have long been recommended as a means of supporting students in developing an understanding of the relations among quantities that underlie mathematical functions. However, there is substantial evidence suggesting that, with current instruction, identifying the link between patterns, generalized rules, and algebraic representations is difficult. Analysts (Lee, 1996; Mason, 1996) who have studied patterns and generalizing suggest that it might not be patterning problems per se that are difficult but the ways that they are presented to students and the limitations of the teaching approaches used. Although patterning problems might be presented initially in a variety of formats (geometric, tabular, and narrative), the tendency of most instruction is to prioritize the numeric/arithmetic aspect of patterning activities (Noss & Hoyles, 1996). This numeric approach to pattern learning diminishes the potential for students to recognize commonalities in mathematical relationships across multiple representations and obscures the underlying functional relationship of the pattern because the pattern rule becomes a sequence of arithmetic operations derived numerically in isolation from the context of the problem. Students do not perceive the need to understand the mathematical structures and relationships underpinning pattern rules, so these activities do little to expand their understanding, ability, or interest in finding and justifying meaningful general rules (functions).

In response to this, Beatty conducted a series of experimental studies in grade 4, 5, and 6 classrooms with the goal of designing new approaches to pattern learning that emphasize multiple representations (geometric, graphic, and symbolic) as a means of supporting children's developing understanding of mathematical functions. Based on Case et al.'s research methodologies (Moss & Case, 1999), the instructional sequences and curriculum materials were aligned with theoretically based learning trajectories and then tested in a variety of classrooms. The inquiry-based lesson sequence was designed to encourage meaning making through the use of mathematical discourse, multiple representations, manipulatives, and rich open-ended tasks. The three-year research program was iterative in that results from each year informed the design of subsequent instruction. Results have demonstrated positive outcomes based on the lesson sequence for this kind of instruction for all students, including those who have demonstrated very low math ability (Beatty & Moss, 2006).

Linking Research, Professional Development, and Teacher Learning

The study for this chapter took place in the third year of the research program. At that point, the content of the lesson sequences had been evaluated, but there still remained questions with respect to implementation of best practices.

Therefore, we invited experienced teachers to participate in the PD program component of the Research/PD Model as collaborative partners to further assess the quality and functioning of our experimental patterning and algebra lesson sequence. Simultaneously, the researchers tested this Research/PD Model to see whether it supported teachers in developing their pedagogical content knowledge, practice, and efficacy (confidence that they can promote student learning) (Bandura, 1977). Teacher efficacy is an important construct because it has been associated with variables such as teachers' willingness to adopt innovations, teachers' classroom management strategies, time dedicated to teaching challenging subjects and topics, and student achievement (Goddard, Hoy, & Woolfolk Hoy, 2004; Ross, 1998; Tschannen-Moran, Woolfolk Hoy, & Hoy, 1998). Fishman's research (Fishman et al., 2003) on enactment details the importance of explicit teacher implementation of strategies modelled in PD sessions in order to positively influence teacher pedagogical content knowledge and efficacy. The pivotal role of teacher enactment—implementation of the content and pedagogy—is central to our model (see Figure 11.1).

This chapter focuses on the shaded part of the model in Figure 11.1 to explore the relationships between implementing a Research/PD Model designed to (1) develop teacher content and pedagogical content knowledge; (2) increase teacher enactment of a new curriculum; and (3) positively influence teacher efficacy.

Research questions for this study were the following.

- Will participation in a PD program based on a specific lesson sequence increase teachers' pedagogical content knowledge in patterning and algebra?
- Will participation in the PD program influence teacher efficacy?
- What is the role of teacher enactment in the professional development of teachers?

The study borrowed from Hill's (2004) model of professional development for PD design and implementation. Hill cites research outlining conditions for effective professional development, stating that "teachers can and do respond to professional development ... when they have adequate opportunities to learn and when those opportunities are focused on content; grounded in artifacts of practice, such as curriculum, videotape, or student work; supported over time ..." (p. 217). Table 11.1 outlines the connection between Hill's recommendations for effective PD design and specific components of our PD program.

Figure 11.1
Research/PD model

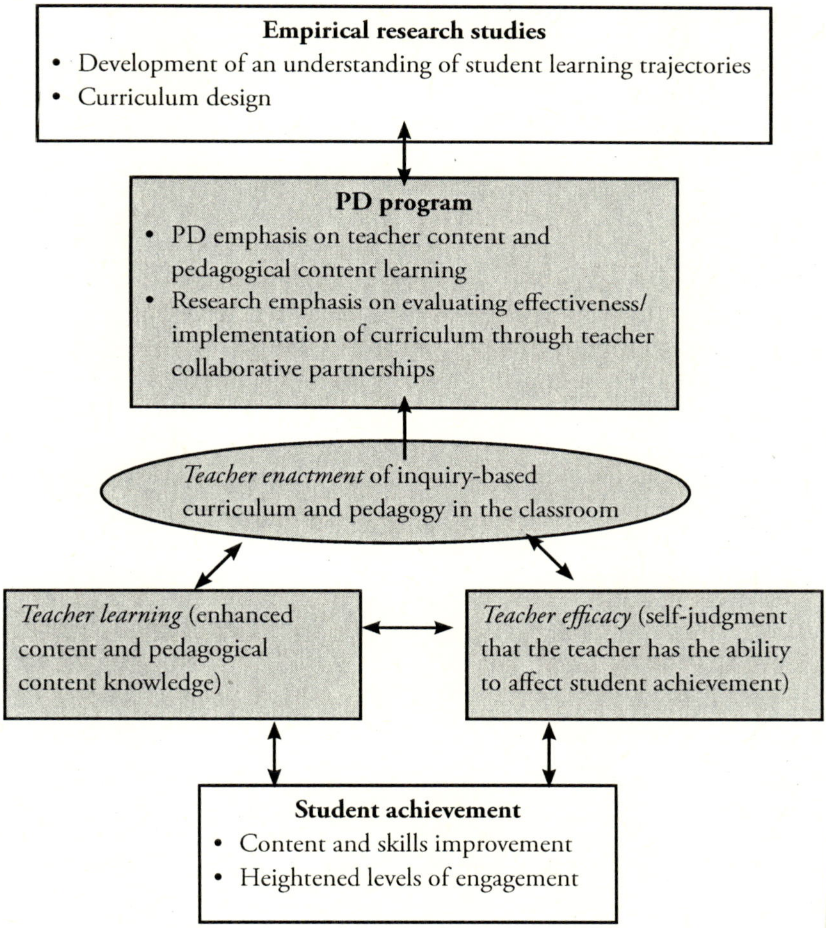

We also used Fishman et al.'s 2003 model to evaluate the outcomes of the PD program, specifically examining the links between teacher participation and enactment in classrooms and changes in teacher knowledge and student outcomes.

Table 11.1
Hill's PD standards matched to our PD program

PD standards	PD components
Active learning: teachers construct meaning through inquiry or analysis to build their own understanding.	Teachers participated in the lesson sequence *as learners,* then implemented it in their own classrooms.
Examples from classroom practice: teachers look at students' work and related curriculum materials	Teachers viewed students' work (artifacts, videotapes, etc.) from the research study. The teachers also focused on student learning in their own classrooms and in the classrooms of other participating teachers.
Collaboration: teachers work together.	Teachers collaborated during the PD sessions and implementation of the lessons.
Modelling: professional developers model constructivist teaching practices.	Researchers modelled not just pedagogy and strategies but also whole sequences of lessons designed to support development of conceptual understanding.
Reflection, practice, feedback: teachers have the opportunity to return to their classrooms, try new ideas, and receive follow-up support.	There were three sessions during which teachers brought back classroom results/experiences to share in focus group sessions.
Focus on content: teachers have opportunities to increase knowledge of the subject matter and pedagogy.	The in-service focus was to develop content knowledge specifically in the domain of patterning and algebra and an understanding of children's thinking in this area.
Focus on student learning: teachers engage in activities that strengthen their understanding of how children learn patterning and algebra.	By engaging in learning and teaching the lesson sequences, teachers were supported in how to understand student thinking.
Teacher choice: teachers choose to participate and make choices based on self-identification of needs.	Participation in this study was voluntary. Teacher choice in terms of subject matter, lessons, etc. was not offered. However, teachers were encouraged to modify lessons (and share modifications) as needed.

Source: Hill, H. C. (2004). Professional development standards and practices in elementary school mathematics. *Elementary School Journal, 104*(3), 215-231.

Methodology

Participants

Fourteen experienced grade 5 teachers and two pre-service teachers from a mid-sized district school board participated in the study.

Five Full-Day PD Sessions (over Four Months)

Each session followed the same format. Each lesson of the sequence was taught to the teachers so that they could develop their own content knowledge about patterns and to model lesson pedagogy. After each lesson, teachers engaged in a researcher-led debriefing session during which the critical learning points were identified and teachers reflected on their experiences with the activity. Then videotape of student learning from the larger project was shown so that the teachers could observe examples of different levels and types of student thinking. The print materials that accompanied the lessons included outlines of the activities, key questions to probe for student understanding, photographs of student artifacts, and sample transcripts of student thinking. We also provided each teacher with a class set of manipulatives.

The initial two-day training session was held prior to implementation of the lesson sequence. During this session, teachers were introduced to the larger patterning and algebra research project and were provided with an overview of the lesson sequence, the developmental learning trajectory, and the underlying rationale for the sequence.

In sessions 2, 3, and 4, which took place during the time period when teachers implemented the lessons in their classrooms, participants debriefed in small focus groups. During these videotaped focus group sessions, led by the board mathematics consultants, the teachers were asked to critique the lessons in terms of content and implementation and with respect to student learning. We were also interested in any modifications the teachers made. Our need for teacher expertise and feedback meant that the teachers were accountable for the implementation of the lessons in their classrooms between PD sessions.

Measures

In this study, we wanted to evaluate the outcomes of teacher participation and examine the effectiveness of the Research/PD Model. Quantitative measures included teacher efficacy pre- and post-surveys (Tschannen-Moran & Woolfolk Hoy, 2001). Qualitative measures included teacher pre- and post-pedagogical content knowledge surveys, transcripts of sixteen focus group interviews, teacher program evaluation forms, and field notes of classroom observations (four per teacher). They were used to critically examine the process of teacher change. Quantitative and qualitative data were analyzed independently in an embedded mixed methods design (Creswell & Plano-Clark, 2007).

Results

This chapter reports the findings related to teacher learning in the form of teacher efficacy and teacher pedagogical content knowledge development.

Teacher Efficacy

Teacher efficacy, the extent to which teachers believe they will be able to bring about student learning, was measured at the beginning and end of the study using twelve items adapted for mathematics teaching from Tschannen-Moran and Wolfolk Hoy (2001). The population sample was small, and therefore reliabilities are less stable than would be the case with a larger sample. Nonetheless, the data gave us one method of determining whether participation in the program had an impact on teachers' confidence in their abilities to affect student learning. There were statistically significant increases in teachers' beliefs about their ability to engage students in mathematics learning, ($t(13) = -1.35$, $p = .02$). An additional twenty items designed to measure attitudes and practices in teaching math indicated that teachers' scores moved toward an inquiry-based stance on the post-test, ($t(13) = -2.18$, $p = .049$).

Focus group interview transcripts and field notes also indicated increased teacher efficacy. For example, Jennifer discovered through the study that "the younger kids can handle a much more complicated understanding level than we thought they could" (Whole Group Transcript, 4.5), suggesting that her estimation of student ability to learn had changed. Grant described his newfound commitment to using innovative teaching strategies once he observed the quality of student learning and engagement:

> I think we really bought into it because we're seeing how the kids would learn. I may have been in a workshop or two that I've just filed, basically I mean I went there, I kind of listened, and I kind of put it away. I don't know, maybe I'm the only one in the board that's done that (laughter). The way that this is laid out—lesson 1, lesson 2, with those key questions—

> I always kept going back, did I ask that question? That's when I really bought into it. (Whole Group Transcript, 4.5)

Andrea described how seeing the students gain confidence increased her confidence to implement new strategies as a mathematics teacher:

> The students have been engaged and have a high level of understanding. They were building patterns, and some were done quick (snaps). I thought, "Some of the kids are done, some aren't, what should I do?" So I asked them to write all the rules for the patterns that they saw, which was really, really good because they all wrote in pairs, and the level of understanding that came out of that blew me away. For some of them, it was phenomenal, it was really good. I would incorporate doing this with other lessons too. They're showing me what they understand, it's really good. (Focus Group Transcript, 3.2)

Teacher Pedagogical Content Knowledge

Teachers' responses to the Pedagogical Content Knowledge Survey were scored using a descriptive scoring guide generated by a math expert. For each question, the scoring guide described three levels of responses: entry level = one mark; mid-level = two marks; and advanced level = three marks, across two criteria: specificity of support for students (i.e., teacher pedagogy), and content knowledge in mathematics (evidenced in the responses). Two teachers did not complete the post-test due to absence, so there were twelve pre- and post-matched teacher responses.

Overall results indicate that there were positive shifts in the twelve teacher responses. On the pre-test, teachers offered general strategies for problem solving (e.g., asking students to explain their thinking, using an ordered table of values). In the post-test, teachers advocated the use of multiple representations, including pattern building, graphing, and finding and applying explicit composite rules. In addition, answers relating to graphs of linear functions showed clear gains in *content* knowledge, for example the relationships between numeric/symbolic and graphic representations such as the coefficient and slope of a line and the constant and *y*-intercept. These results suggest that the PD program might have had a positive effect on teacher pedagogical and content knowledge in mathematics related to patterning and linear functions.

Teachers also reported increased confidence in adopting a student-directed approach that included rich classroom discussion, use of manipulatives, and

Table 11.2
Results of pre- and post-comparisons of pedagogical content knowledge

	Pre-means (out of 3.0)	Post-means (out of 3.0)
Specificity of support for students (pedagogy)	1.3	1.6
Math content knowledge evident	1.3	1.9

open-ended tasks. The pedagogical content knowledge learned through implementing the lesson sequence generalized to other math domains:

> So I am comparing teaching algebra to fractions because the kids are so successful in the patterning work and able to explain it. ... If I can use the same approach with fractions, getting out the manipulatives and doing the group work and the modelling, maybe they'll have the same foundation for fractions, ... laying down the conceptual base so students have a solid understanding. (Field Notes, 4.2).

Teacher engagement in mathematics teaching and ownership of the lesson sequence increased over the duration of the study based on classroom observation data (Observations, 1-51). Teachers attributed these increases to (in order of teacher-ranked importance) quality of the lesson materials provided, opportunities to practise and discuss the lessons during PD sessions, and enacting/implementing the lesson sequence in the classroom (Teacher Evaluation Forms).

Discussion

The findings have led to a confirmation of the hypothesized model of Figure 11.1 for this study. We found that teacher engagement in high-quality PD (Hill, 2004), combined with accountability for enactment of the proposed lesson sequence, resulted in increased teacher confidence to take risks with innovative teaching strategies and increased expectations of their abilities to positively influence student learning.

The results clearly indicate that the Research/PD Model had a positive impact on teacher learning. By charging teachers with the role and responsibility of implementing the lesson sequence and critiquing its effectiveness, teachers increased their own understanding of, and success with, teaching patterning and algebra while their students also gained understanding.

Although our primary goal was not to examine student outcomes, we recognize the fundamental link between teacher learning and student outcomes (for an analysis, see Beatty, Bruce, & Chahine, 2008). Teachers developed a sufficient level of content and pedagogical content knowledge so that positive student outcomes were comparable to those reported in our previous studies with smaller populations. We believe that this model, in which researchers work alongside teachers, strengthens the link between empirical classroom-based research and successful broader classroom implementation.

Student mastery experiences enhanced teachers' feelings of efficacy as the study progressed. This in turn increased teachers' commitment to implementing the lessons and participating in a professional collaborative learning experience. Teachers reported that changes in their practice extended beyond patterning and algebra to recognition of the importance of developing conceptual mathematical understanding through the use of an inquiry-based approach.

In summary, teachers began to conceptualize the teaching and learning of mathematics as a dynamic, creative endeavour both for students—"They are so engaged, and every single group was focused on the mathematics, which we know doesn't always happen in the classroom" (Focus Group, 3.2)—and for themselves:

> And I'm really thinking that this is what math is all about, even as a teacher to get to the point where you can't wait to teach these math classes because they are so much fun and because there's a point where you *know* one of the students is going to do something or say something really incredible and *out there,* and you're going to go "What? You are opening up a little part of my brain!" And the more you see your kids building confidence, the more confidence you have as a math teacher. (Focus Group, 3.4)

Sustainability of shifts in teacher practice and conceptions of math teaching and learning is promising, as all of the teachers in the study have voluntarily continued their professional development in self-directed study groups with support from the board consultants and researchers. "I just—I want to know we can continue to do this, meet and discuss like this and keep learning deeply about math teaching. I feel like I need to keep learning and growing with everyone, with all of us thinking about this and talking it through. This is the most important thing for me ..." (Focus Group, 4.4).

References

Bandura, A. (1977). Self-efficacy: Toward a unifying theory of behavioural change. *Psychological Review, 84,* 191-215.

Beatty, R., Bruce, C., & Chahine, S. (2008). Evaluating a research/professional development model through student outcomes. Paper presented at the annual meeting of the American Educational Research Association, March.

Beatty, R., & Moss, J. (2006). Multiple vs. numeric approaches to developing functional understanding through patterns: Affordances and limitations for grade 4 students. In S. Altorre, J. L. Cortina, M. Saiz, & A. Mendez (Eds.), *Proceedings of the twenty-eighth annual meeting of the Psychology of Mathematics Education, North American Chapter,* vol. 2 (pp. 87-94). Merida, Mexico: PME-NA.

Creswell, J. W., & Plano-Clark, V. L. (2007). *Designing and conducting mixed methods research.* Thousand Oaks, CA: Sage.

Fishman, B. J., Marx, R. W., Best, S., & Tal, R. T. (2003). Linking teacher and student learning to improve professional development in systemic reform. *Teaching and Teacher Education, 19,* 643–658.

Fullan, M., Hill, P., & Crevola, C. (2006). *Breakthrough.* Thousand Oaks, CA: Corwin Press.

Goddard, R. D., Hoy, W. K., & Woolfolk Hoy, A. (2004). Collective efficacy beliefs: Theoretical developments, empirical evidence, and future directions. *Educational Researcher, 33*(3), 3-13.

Hill, H. C. (2004). Professional development standards and practices in elementary school mathematics. *Elementary School Journal, 104*(3), 215-231.

Lee, L. (1996). An initiation into algebraic culture through generalization activities. In N. Bednarz, C. Kieran, & L. Lee (Eds.), *Approaches to algebra: Perspectives for research and teaching* (pp. 87–106). Dordrecht: Kluwer.

Mason, J. (1996). Expressing generality and roots of algebra. In N. Bednarz, C. Kieran, & L. Lee (Eds.), *Approaches to algebra: Perspectives for research and teaching* (pp. 65–86). Dordrecht: Kluwer.

Moss, J., & Case, R. (1999). Developing children's understanding of the rational numbers: A new model and an experimental curriculum. *Journal for Research in Mathematics Education, 30*(2), 122-147.

Noss, R., & Hoyles, C. (1996). *Windows on mathematical meanings: Learning cultures and computers.* Dordrecht: Kluwer.

Ross, J. A. (1998). The antecedents and consequences of teacher efficacy. In J. Brophy (Ed.), *Research on teaching,* vol. 7 (pp. 49-74). Greenwich, CT: JAI Press.

Tschannen-Moran, M., & Woolfolk Hoy, A. (2001). Teacher efficacy: Capturing an elusive construct. *Teaching and Teacher Education, 17,* 783-805.

Tschannen-Moran, M., Woolfolk Hoy, A., & Hoy, W. K. (1998). Teacher efficacy: Its meaning and measure. *Review of Educational Research, 68*(2), 202-248.

Chapter 12

Professional Development: From Telling to Listening

Christine Suurtamm and Nancy Vézina

Our study reports on a professional development initiative designed to provide teachers with experiences to support them as they move from a model of teacher explanation and repeated student practice to attending to students' mathematical thinking within problem-solving settings. Such a shift in teacher practice is challenging and difficult to put into action. This study, which included 147 teachers, helps to shed light on the professional development experiences that support teacher learning.

Current Perspectives on Mathematics Teaching and Professional Development

Current perspectives in mathematics education focus on the understanding of mathematical concepts, procedures, connections, and applications through problem solving (National Council of Teachers of Mathematics [NCTM], 1989, 2000; National Research Council, 1989). As such, meaningful mathematics takes place in classrooms that help students to investigate, represent, and connect mathematical ideas through mathematical discussions in the context of problem solving (Hiebert et al., 1996; Lampert, 1990; NCTM, 1989, 2000). However, facilitating such problem solving and mathematical inquiry poses substantial challenges for teachers. The posing of problems, the facilitation of discussion, and the consolidation of mathematical concepts are teacher practices that require a great deal of expertise and attention. Teachers struggle with issues such as when to explain and when to listen, how much time to leave for ideas to emerge, and which question to ask next (Suurtamm & Graves, 2006).

Several reasons have been suggested for why this shift in practice is challenging. For example, in many cases, teachers themselves have not learned mathematics in this way, nor have they had opportunities to learn or teach in inquiry-oriented settings. New ways of thinking about mathematics teaching require new understandings among teachers and involve "observing students, listening carefully to their ideas and explanations, having mathematical goals, and using the information to make instructional decisions" (NCTM, 2000, p. 19). Additionally, focusing on students' thinking is challenging as students present various ways of thinking about a mathematical problem (Ball, 2001; Schifter, 2001). For elementary teachers, a limited understanding of mathematics can make them hesitant to engage students in open-ended problem solving or prevent them from recognizing and furthering the important concepts inherent in mathematical activity (Ball, 1990, 1999; Ball, Lubienski, & Mewborn, 2001). Working with teachers to learn to pose problems, to provide opportunities for students to explore the problems, and to listen to their solutions takes time and poses several challenges. Teachers often perceive their roles as explaining mathematical facts and procedures. Professional development, therefore, has the challenge of shifting these perceptions as well as addressing new mathematics content and pedagogies.

Although there is no perfect model for professional development, several characteristics of effective professional development for mathematics teachers have been highlighted in the literature (Loucks-Horsley, Stiles, & Hewson, 1996; Mundry, 2005; NCTM, 1991). Professional development should present a vision of a mathematics classroom that rests on research about how children learn and provide a window into the nature of mathematical teaching and learning. Participants should be engaged in mathematical content as well as pedagogy (Ball, 2003; Cohen & Hill, 2000; Crespo, 2000). Strategies, such as investigation and collaborative work, that serve as models for effective teaching and learning should also be used in professional development. Research has also shown that professional development that is sustained over time (Carroll, Fuson, & Diamond, 2000) and connected with the teachers' own classroom experiences supports new teacher learning (Bednarz, 2000). Research also indicates the importance of attending to individual teacher needs, providing learning opportunities tailored to those needs, and creating a climate of collegiality and support (Mewborn, 2003).

Context of Our Study

Our research focuses on a particular professional development initiative set within the community of French-language educators in the province of Ontario, Canada. The francophone community represents 4.8% of the population of Ontario and has its own set of publicly funded school boards. Although curriculum documents, set by the provincial government, are provided in both French and English, French-language resources that align with the curriculum are lacking. Thus, this project began with funding for a group of French-language mathematics educators from various school boards to develop pedagogical resources for teaching mathematics in the French-language schools. This initiative evolved to include field-testing the resources with francophone teachers from different grade levels and to support them as they used the resources. The team of resource developers thus also became a professional development team that we will refer to as the Resource/Professional Development (RPD) team.

The Professional Development Initiative

The professional development (PD) initiative extended over a three-year period. The first year focused on the development of primary mathematics and included kindergarten to grade 3 teachers; the second year was aimed at teachers of grades 4 and 5; and the third year focused on grades 6, 7, and 8. Each year followed a similar pattern and engaged the group of teachers for one full school year, with PD meetings spread throughout the year so that teachers had the opportunity to try new ideas in their classrooms and discuss them with their colleagues. The PD sessions focused primarily on teachers' understanding of mathematics and on ways to help students develop mathematical understanding. In the first meeting, the RPD team shared examples of current research in math education to provide a rationale for changing some of the traditional approaches to mathematics so that students could develop a better understanding of mathematical concepts (Carpenter, Ansell, Franke, Fennema, & Weisbeck, 1993; Cobb et al., 1991). Teachers discussed topics such as teaching through problem solving and connecting school mathematics with the mathematics in students' lives. All sessions engaged the participants in mathematical problem solving and solution sharing, working with manipulatives, viewing videos of classroom activities, and engaging in discussions. These activities focused on strategies to help students develop mathematical understanding. Time was also devoted to sharing experiences of using the new resources in classrooms. As well as formal professional development meetings at different times of the year, the teachers were also provided with assistance from their math consultants through visits to their classrooms. These visits took a variety of forms, such as observation, co-teaching, or assistance with preparation of materials. The teachers were also able to contact the RPD team directly if they had questions or needed assistance.

Participants

In this chapter, we include the work from all three years of the project. This includes twenty-three kindergarten through grade 3 teachers from the first year; thirty-nine grade 4 and 5 teachers from the second year; and eighty-five grade 6, 7, and 8 teachers from the third year. The teachers participated in school groups; for example, a school would send all the kindergarten to grade 3 teachers from that school to participate in the initiative. The mathematics consultants from the school boards involved in the project also received the training so that they knew what the teachers were learning, and they supported the teachers through their visitations, observations, or assistance in classrooms in the schools. The teachers had a variety of experiences ranging from a few beginning teachers to over 50% with more than nine years of teaching experience. The participants in this PD initiative—classroom teachers, mathematics consultants, and the RPD team—were also participants in our research study. Our involvement in this initiative was solely as researchers. In other words, the professional development was designed and carried out by the RPD team, which did not include us.

The Study

This project is a specific example of a group of teachers involved in a unique initiative to learn more about teaching mathematics within a supportive environment. As Merriam and Simpson (1995) suggest, a case study approach helps to provide an intense description of a group or community to illustrate the interplay of various factors and the complexity of human situations. In examining teachers' experiences in this situation, we draw on a social constructivist framework (Confrey, 1990; Davis, Maher, & Noddings, 1990; Forman, 2003). From this perspective, new understanding is constructed in relational networks that emerge from the interactions of people and activity environments. This shifts the traditional view of learning from acquisition to participation (Sfard, 1998), important for conceptualizing professional development as situations in which to build communities for inquiry into practice.

To answer our research questions and to gather data from the multiple components of the professional development initiative, we used a case study design that included several sources of data. For each year of the project, two questionnaires were administered. The first was at the beginning of the school year, and the second was halfway through the year. The first questionnaire gathered information from the teacher about his or her current and past teaching experience, level of confidence about mathematics teaching, and reasons for involvement in the project. The second questionnaire gathered data on teachers' classroom experiences and ways that they were supported and challenged. Data were also collected through focus group interviews with participating teachers, consultants, and the RPD team. All interviews were audio-recorded and transcribed. We also had access to training materials and the full pedagogical resource package created for each grade. A qualitative analysis of the data was completed through content analysis of the open-ended questions from the teachers' questionnaires and the transcribed discussions of the focus groups as well as an analysis of the professional development sessions and videotaped classroom experiences. Our analysis was inductive to uncover emergent categories, themes, and concepts (Guba & Lincoln, 1981).

Results

Both the teacher questionnaires and the teacher focus group data provide several themes that include teachers' enhanced understanding of mathematics, changes in their classroom practices, students' increased understanding of mathematics, and particular challenges. These themes were reinforced by the analysis of the focus group data from the mathematics consultants and the RPD team who were part of the PD sessions and observers of participants' classrooms.

Teachers reported an increase in their own understanding of mathematics and suggested that working with different methods and strategies allowed them to understand some of the standard algorithms that they had been using for years. "I thought that I understood numeration, such as work with powers of 10, but I realize that I had just memorized when you add or remove a 0" (grade 4-5 teacher, Focus Group H, June 2006). Many expressed the wish that they had been taught using these methods. They found that they now enjoyed doing mathematics, a change from their previous attitude. Although some mentioned that they sometimes have difficulty setting aside the traditional algorithms when they themselves are doing mathematics, there were moments when they found themselves using some of the invented strategies that their students used. For instance, one grade 4 teacher described a scenario of working with a colleague and coming to a point where she needed to multiply multi-digit numbers. She realized that, rather than reaching for her calculator, she could use an area model to do the multiplication, as students in her class had done. She produced the accurate result, and although her colleague reached for the calculator to check the answer she had no reason to doubt her own accuracy. She reflected on what a change this was for her in her approach to solving problems.

Teachers discussed many changes that they were making in their classroom practice. They worked on developing particular problem-solving practices such as presenting students with a problem, giving them time to work in pairs to generate multiple solutions, and then sharing solutions in small group and whole class settings. But along with such structures, they found that they increased questioning, promoted more communication, encouraged greater use of manipulatives, and were more attentive to student solutions. In terms of questioning, most teachers said that they were using questioning more effectively and encouraging multiple ways of looking at a task or solving a problem. The teachers promoted a different learning environment, encouraging risk taking and more communication, specifically more mathematical talk between students. "Yes, it has improved our students' communication. I think it was where I needed to improve as a teacher—to help my students communicate more. They're anxious to share and participate now. They go to the board in pairs and explain what they did" (grade 4–5 teacher, Focus Group G, June 2006). They noted that increased student communication made them more aware of and responsive to students' thinking.

> I think that my greatest learning is to be able to know where the student is in his learning process. When I went back to the classroom, I could see where my students were. Some of them were far ahead in the learning continuum, and others were at the beginning, and I was able to change

> my teaching based on recognizing where my students were. (grade K–3 teacher, Focus Group D, June 2005)

Teachers also reported that they increased their use of concrete materials in a more meaningful way, moving from manipulatives as fun to using them as representations of mathematical ideas. They also saw that manipulatives were beneficial to all, not just struggling students. "It is important to use manipulatives with the older students. ... They are not only for primary grades. We need to use them in kindergarten, grade 1, grade 2, grade 3, and so on. Continuity is important" (grade K–3 teacher, Focus Group A, June 2005). Most importantly, teachers developed ways of listening and responding to student solutions and recognized that students are able to correct their own errors and develop new ways of solving problems through the sharing of solutions.

Teachers also overwhelmingly reported changes in their students' learning and attitude toward mathematics. Students seemed more positive, mathematics seemed more interesting to them, and they were involved in more mathematical activity. Teachers particularly noticed that students who had struggled with mathematics were having more success. "Even students that used to have difficulties want to share as well. For instance, one of my students who used to struggle has changed completely. He now wants to get involved, go to the board, and explain his thinking. Sometimes he's the only one who can figure out a problem" (grade 4–5 teacher, Focus Group G, June 2006). Teachers thought that all students were now engaged in exploring mathematical ideas and thought that their ideas were valued.

Teachers also discussed several challenges in changing their practice. As one teacher reported, "Your brain has been used to doing it the old way for twenty years." The teachers of grade 3 were concerned whether their students would be prepared for the large-scale provincial assessment given at the end of that grade. Teachers were also concerned whether the teachers whom the students would have in the following years would value the different ways of reasoning about mathematics or whether they would assume that students should know the standard algorithms and procedures. Teachers also recognized that, although they had support for their teaching from their colleagues and the school board consultant, support from the principal and parents was also a necessary element. The principal was often the gatekeeper to needed resources and the one who would need to communicate with parents about some of the changes in teaching practice.

Discussion

Our research provides evidence that the teachers in this study have developed a deeper understanding of the mathematics that they teach, are expanding their pedagogical practices to encourage students to engage in problem-solving activities, have created a classroom culture in which students share their mathematical ideas, and have learned the value of listening to students' thinking as a means to help both students and teachers develop new understandings. We now would like to turn our discussion to some of the factors that have helped to facilitate these new understandings and practices. Our data analysis demonstrates that the professional development initiative described in this study had many of the components suggested by the literature discussed at the beginning of this chapter. For instance, the RPD team shared a vision of mathematics teaching and learning that incorporated current research through their use of videos of classroom activities and prepared resources. The initiative engaged teachers and consultants in mathematical content and pedagogy through problem-solving activities that included posing questions, using manipulatives, and sharing solutions and strategies. Such activities also modelled investigative and collaborative work and created a community of learners among the teachers that modelled the type of community that teachers could create in their classes. The PD initiative was also sustained over time as teachers were involved over a year and moved back and forth from professional development sessions to their classrooms, which helped them to test new ideas and connect theory and practice. But, most importantly, all of these aspects helped to create a climate of collegiality and collaboration in which teachers felt comfortable trying out and discussing new ideas, and it is this collaborative and collegial environment that we would like to highlight in more detail.

The teachers reported that collaboration played a major role in creating a community to support their work, and this collaboration took on a variety of forms. The PD initiative worked with groups of teachers from the same schools. This created a collaborative community, both within the professional development experience and back at the school. Thus, when teachers were trying out new ideas, they had someone in the classroom next to them with whom to share their dilemmas, successes, and challenges. As an example, one teacher mentioned that previously, when she saw a student solution that she did not understand, she would tend to mark it wrong. She reported that now she brought the solution to the teacher next door, and together they understood the student's thinking. Collaboration also occurred between the teachers and the consultants as the visits from the consultants not only provided feedback but also served as an opportunity for the teachers to voice their concerns or discuss issues that they had in their classrooms. This helped to create a collaborative environment that connected with someone outside the school but still focused on the classroom context and provided support to the teacher. Furthermore, collaboration occurred at the professional development meetings as teachers connected with teachers and consultants from other schools and districts. Teachers also had strong relationships with the RPD team as they shared their classroom experiences and provided feedback to the RPD team about the resources that they were developing. This collaboration provided teachers with new ideas to take back to the classrooms and gave the RPD team valuable feedback to incorporate into their resource development and future PD meetings. Thus, the collaboration was useful to all members of the initiative.

Concluding Comments

In summary, the professional development initiative mirrored the learning and teaching environment that the RPD team was encouraging the teachers to generate in their classrooms. The initiative created a collaborative and safe environment for risk taking so that teachers could try out new ideas in much the same way that they were creating safe environments for their students to try out a variety of strategies in problem solving. The teachers were encouraged to communicate their new understandings in ways similar to those that they had learned to encourage students to communicate their different strategies for solving problems. The teachers also saw the value in learning from one another and from the students in much the same way that students saw the value of learning from other students' strategies. But, most importantly, the teachers listened and were listened to as they developed new pedagogical approaches and the skill of listening in their own classes. As one teacher summarized, "I think I can compare how kids feel to what we have done because we felt the same way. Because we had the experience of doing this ourselves in the training, our confidence to proceed was based on experience rather than theory" (grade 4–5 teacher, Focus Group H, June 2006). This quotation reflects the environment that the teachers experienced as they learned and tried out new pedagogical and mathematical understandings. Our findings support the notion that the model of teachers moving from telling to listening in their classrooms also served as a sound professional development model.

References

Ball, D. L. (1990). The mathematical understandings that prospective teachers bring to teacher education. *Elementary School Journal, 90*(4), 449-466.

Ball, D. L. (1999). Crossing boundaries to examine the mathematics entailed in elementary teaching. In T. Lam (Ed.), *Contemporary mathematics* (pp. 15-36). Providence, RI: American Mathematical Society.

Ball, D. L. (2001). Teaching, with respect to mathematics and students. In T. Wood, B. S. Nelson, & J. Warfield (Eds.), *Beyond classical pedagogy: Teaching elementary school mathematics* (pp. 11-22). Mahwah, NJ: Lawrence Erlbaum Associates.

Ball, D. L. (Ed.). (2003). *Mathematical proficiency for all students: Toward a strategic research and development program in mathematics education.* Santa Monica: RAND Institute.

Ball, D. L., Lubienski, S., & Mewborn, D. (2001). Research on teaching mathematics: The unsolved problem of teachers' mathematical knowledge. In V. Richardson (Ed.), *Handbook of research on teaching,* 4th ed. (pp. 433-456). New York: Macmillan.

Bednarz, N. (2000). Formation continue des enseignants en mathématiques: Une nécessaire prise en compte du contexte. In P. Blouin et L. Gattuso (Dir.), *Didactique des mathématiques et formation des enseignants* (pp. 63-78). Mont-Royal, QC: Modulo Éditeur.

Carpenter, T. P., Ansell, E., Franke, M. L., Fennema, E., & Weisbeck, L. (1993). Models of problem solving: A study of kindergarten children's problem-solving processes. *Journal for Research in Mathematics Education, 24*(5), 427-440.

Carroll, W., Fuson, K., & Diamond, A. (2000). Use of student-constructed number stories in a reform-based curriculum. *Journal of Mathematical Behavior, 19*(1), 49-62.

Cobb, P., Wood, E., Yackel, J., Nicholls, G., Wheatley, B., Trigatti, B., et al. (1991). Assessment of a problem-centered second-grade mathematics project. *Journal for Research in Mathematics Education, 22*(1), 3-29.

Cohen, D. K., & Hill, H. C. (2000). Instructional policy and classroom performance: The mathematics reform in California. *Teachers College Record, 102,* 294-343.

Confrey, J. (1990). What constructivism implies for teaching. In R. B. Davis, C. A. Maher, & N. Noddings (Eds.), *Constructivist views of the teaching and learning of mathematics, Journal for Research in Mathematics Education* Monograph 4 (pp. 107-122). Reston, VA: National Council of Teachers of Mathematics.

Crespo, S. (2000). Seeing more than right and wrong answers: Prospective teachers' interpretations of students' mathematical work. *Journal of Mathematics Teacher Education, 3*(2), 155-181.

Davis, R. B., Maher, C. A., & Noddings, N. (1990). Introduction: Constructivist views on the teaching and learning of mathematics. In R. B. Davis, C. A. Maher, & N. Noddings (Eds.), *Constructivist views of the teaching and learning of mathematics, Journal for Research in Mathematics Education* Monograph 4 (pp. 1-3). Reston, VA: National Council of Teachers of Mathematics.

Forman, E. A. (2003). A sociocultural approach to mathematics reform: Speaking, inscribing, and doing mathematics within communities of practice. In J. Kilpatrick, W. G. Martin, & D. Schifter (Eds.), *A research companion to principles and standards for school mathematics* (pp. 333-352). Reston, VA: National Council of Teachers of Mathematics.

Guba, E. G., & Lincoln, Y. S. (1981). *Effective evaluation.* San Francisco: Jossey-Bass.

Hiebert, J., Carpenter, T. P., Fennema, E., Fuson, K., Human, P., Murray, H., Olivier, A., & Wearne, D. (1996). Problem solving as a basis for reform in curriculum and instruction: The case of mathematics. *Educational Researcher, 25*(4), 12-21.

Lampert, M. (1990). Connecting inventions with conventions. In L. P. Steffe & T. Wood (Eds.), *Transforming children's mathematics education: International perspectives* (pp. 253-265). Hillsdale, NJ: Erlbaum.

Loucks-Horsley, S., Stiles, K., & Hewson, P. (1996). Principles of effective professional development for mathematics and science education: A synthesis of standards. *National Institute for Science Education Brief, 1*(1), 1-6.

Merriam, S. B., & Simpson, E. L. (1995). *A guide to research for educators and trainers of adults,* 2nd ed. Malabar, FL: Krieger Publishing.

Mewborn, D. S. (2003). Teaching, teachers' knowledge, and their professional development. In J. Kilpatrick, W. G. Martin, & D. Schifter (Eds.), *A research companion to principles and standards for school mathematics* (pp. 45-52). Reston, VA: National Council of Teachers of Mathematics.

Mundry, S. (2005). *What experience has taught us about professional development.* Eisenhower Mathematics and Science Consortia and Clearinghouse Network, retrieved 20 October 2007 from http://www.sedl.org/pubs/catalog/items/ms90.html.

National Council of Teachers of Mathematics (NCTM). (1989). *Curriculum and evaluation standards for school mathematics.* Reston, VA: NCTM.

National Council of Teachers of Mathematics (NCTM). (1991). *Professional standards for teaching mathematics.* Reston, VA: NCTM.

National Council of Teachers of Mathematics (NCTM). (2000). *Principles and standards for school mathematics.* Reston, VA: NCTM.

National Research Council (NRC). (1989). *Everybody counts: A report to the nation on the future of mathematics teaching.* Washington, DC: National Academy Press.

Schifter, D. (2001). Learning to see the invisible: What skills and knowledge are needed to engage with students' mathematical ideas? In T. Wood, B. S. Nelson, & J. Warfield (Eds.), *Beyond classical pedagogy: Teaching elementary school mathematics* (pp. 109-134). Mahwah, NJ: Lawrence Erlbaum Associates.

Sfard, A. (1998). On two metaphors for learning and the dangers of choosing just one. *Educational Researcher, 27*(2), 4-13.

Suurtamm, C., & Graves, B. (2006). Immersion in mathematical inquiry: The experiences of beginning teachers. In J. Novotná, H. Moraová, M. Krátká, & N. Stehlíková (Eds.), *Proceedings of the 30th conference of the International Group for the Psychology of Mathematics Education,* vol. 1 (p. 340). Prague: PME.

Chapter 13

Continuing Professional Education of Mathematics Teachers: Learning by Investigating[1]

Cármen Lúcia Brancaglion Passos

and Maiza Lamonato

Introduction

This research is based, on the one hand, on the relevance of considering the teaching of geometry at the preschool level and, on the other, on the potentiality of considering mathematical investigation for the teaching and learning of mathematics. It is also founded on the fact that teachers are the central agents of their professional development and construct professional knowledge throughout their practice. Consequently, the learning revealed by these teachers who teach mathematics to six-year-old children, during the Continuing Professional Training Course, is central in the research study.

Teachers' Knowledge

Understanding that teachers' knowledge development is a continuing process, which includes pre-service education, we consider also that it should be based on reflective daily activity to improve knowledge, understanding, and teaching skills, so that awareness of that development can act as a self-trigger (Nacarato & Paiva, 2006; Ponte, 1994). Therefore, it is pertinent to understand how teachers perceive knowledge required and used in teaching:

> [This knowledge is a] reflexive knowledge, with plural and complex historical reasons, [it is] provisory, contextual, affective, ethical-political [its underlying work objectives are human beings], [it is] cultural, forming a network, more or less coherent and entrenched, of scientific knowledge—rooted in science education, in the specific content knowledge, in the content curriculum—and in experienced knowledge as well as in pedagogical tradition. (Fiorentini, Nacarato, & Pinto, 1999, as cited in Fiorentini, 2000, p. 190).

As explained by Saraiva and Ponte (2003), citing Serrazina (1998), the reflection done by a teacher on his or her experience is essential to analyzing the assumptions that give meaning to teaching practices and therefore to improving his or her practices and deepen his or her understanding.

So, according to the theoretical evidence presented in the next section as well as the methodological choices that we made, in conceiving of the possible learning that a teacher does in a teaching-learning process, we consider that a mathematical investigation can be a trigger to discussion, problematization, and enlargement of teachers' knowledge.

Mathematical Investigation and Geometrical Teaching

The mathematical investigations used during the Continuing Professional Training Course were designed on the basis of the characteristics of investigation classrooms such as defined by Fiorentini (2006, p. 29):

> Those that trigger, in classroom settings, tasks and open activities, exploratory situations and not guidelines to the students' thought and that have multiple approaches, strategies and solutions. ... Depending on the way these classes are developed, the activity can be restricted to an exploratory and problematization stage. However, if during the development of the activity, questions and conjectures are made that can trigger a testing process, where attempts are made to demonstrate or prove those conjectures, then we have a mathematical investigation.

In a mathematical investigation, those who have developed the activity, through a critical observation of a given situation, make the presentation of a problem. As Freire and Faundez (1985) state, knowledge begins with a question. Since the ability to conceive of questions is the foundation of knowledge, it is considered as the first mathematical activity. A mathematical investigation in a classroom tends to be compared with the mathematical activity of professional mathematicians (Abrantes, Leal, & Ponte, 1996; Ernest, 1996), particularly in relation to the nature of the activities and the attitudes involved in the two cases, such as curiosity, questioning, formulation of hypotheses, elaboration and testing of conjectures, justification, reasoning, and discussion of the work conducted.

In the case of geometry, Abrantes (1999) restates the link between this mathematical domain and mathematical investigation, which allows the discovery of new concepts and problem-solving strategies through an emphasis on intuition and visualization. These types of activity do not require a huge number of prerequisites and contribute to a mathematical development that goes beyond algorithms and recipes.

Teaching geometry so that students can understand space, position, and localization and develop perception and orientation skills is a key component in their understanding of, adapting to, and exploring the world around them. This is why geometry is so important at all the teaching levels.

Methodology

This qualitative research involved four teachers—Laura, Guilhermina, Ana Júlia, and Bianca—teaching in public schools in an average-size city in São Paulo state, Brazil.

Data were collected through the researcher's diaries, written records of teachers' productions, and video-recordings of the sessions of a Continuing Professional Training Course aimed at teachers teaching mathematics to six-year-old children. In this course, the researcher acted as a participant-observer; the observer was here part of the situation observed "through a direct and personal observation" (Laville & Dionne, 1999, p. 180).

In the Continuing Professional Training Course, several geometrical investigation tasks were proposed to teachers. Along with these tasks, teachers had to discuss aspects related to their pre-service and in-service education as well as to their own teaching practices. Moreover, teachers implemented investigative tasks in their mathematics teaching for six-year-old children, designing possible tasks and developing investigative activities in their classrooms.

An analysis of the data was conducted, taking account of different eleme nts related to teachers' knowledge, teaching of geometry, and mathematical investigation in mathematics teaching. This analysis tried to answer the following main research question. *What knowledge is revealed by pre-school teachers when discussing geometry and its teaching in an investigative context?*

In this chapter, we focus more on aspects such as *the rethinking of teaching practices and the reconstruction of teaching knowledge.*

Rethinking Teaching Practices, Reconstructing Teaching Knowledge

The data collected reveal that it was important, during the teachers' education course, to provide an environment in which teachers could construct their own knowledge in relation to a certain mathematical content knowledge. Teachers had to perceive that production as a creative action allows them to establish a new relation to mathematics, "calling for imagination, intuition, emotions, the ability to make conjectures, to compare, to accept an hypothesis until an explication is reached" (Freire, 1996, p. 88).

One of the dialogues recorded supports the importance of such an environment. The teachers pointed out the attitudes of the teacher educators, who did not give them ready answers but pushed them to find out, to think of, the solutions. The teachers explained that this approach is also desirable when working with children.

Bianca: I was provoking you. ... The way you led things. ...
Guilhermina: She made you prove!
Bianca: Not in a direct way, talking and explaining, ... but in a way that we were pushed ... to find out.
Laura: Pushing us, in fact.
...
Laura: To think!
Guilhermina: That is what we are supposed to do with children.
Laura: She didn't give us a straight answer at any moment.

This possibility pointed out by the teachers—in which the tasks are open and the teacher pushes students to make discoveries and justify them—became the main concern during the course. It became a challenge for their own practice. Guilhermina presented her own discontent with her teaching: she considered that she overdirected her students instead of allowing them to discuss, to think, and to come up with different answers: "I can't think of an activity. ... I am not happy. ... I don't know how not to lead the pupils, how to let them think, discuss, and come up with different answers."

The activities in mathematics teaching in which pupils can discuss, argue, have different answers, reach new discoveries, and not have ready replies can be put in relation to the meaning of the mathematical investigations in which the teachers were involved during the period when the data were collected.

At a later date, Guilhermina conducted an investigative activity with her students (six-year-old pupils). With a set of packages, the children discussed and decided how they could make a toy, the *bilboquet.*[2] In her report, we can see that the children raised hypotheses and tested them to decide which of the packages could be used to make the toy, and they made decisions connected to the proper size of the toy. Instead of doing what she used to do, Guilhermina did not lead the pupils or give them ready answers so that they would reach the same answer that she did. Instead, she gave them chances to raise, discuss, and test their hypotheses and justify their choices.

> I liked it because they were able to raise hypotheses. When I proposed the task, I placed three types of packages: a box of soap, a water bottle, and a toilet paper roll. They had to explain their choice. Then they had to raise a hypothesis and justify it. Then we did it. Afterward they had to raise a hypothesis regarding the size of the toy. Again they raised a hypothesis. ... The good thing was that they could actually see if their hypothesis was real or not.

Guilhermina's dynamics in this classroom reveal ways of investigating tasks in preschool teaching: *the teacher was searching with the pupils.* This type of knowledge was developed by the teachers in the Continuing Professional Education Course, in which they experienced this type of activity and its advantages, making them want to include it in their teaching practices.

Beyond the development revealed by the teachers in their teaching practices, the analysis pointed to specific content learning in their reflections and worries related to their teaching skills and practices, as we can see in the dialogue between Guilhermina and Laura. These teachers considered squares and rectangles as polygons that belong to different groups: that is, they did not perceive a square as a rectangle. After doing one of the activities of the course, Guilhermina revealed her doubt about teaching these concepts: "When we saw it there, 'the square is a rectangle,' then I said, 'How am I going to differentiate them for my pupils?'"

As Mizukami (2004) states, it is not enough for teachers to have intuitive or personal knowledge of a concept or to have principles or particular theories. A teacher should know the ways of transforming these contents according to the purpose of teaching. Moreover, a teacher has to be coherent in terms not only of the knowledge related to a particular content in a specific area, including facts, concepts, and procedures, but also of how this knowledge can be constructed and validated within the specific area (Mizukami, 2004; Shulman, 1986).

Conclusion

The data analysis suggests that the mathematical investigations developed by the teachers during the course enabled them to reconstruct knowledge related to geometrical content as well as to restructure their role and action in the classroom. This new role can enable pupils to construct their own knowledge by investigating, questioning, raising hypotheses, and making conjectures and validating them, together with the discussion and argumentation presented in relation to the ideas and solutions developed. These processes are similar to those of a professional mathematical activity and diverge from the lecturing and learning of ready-made ideas, solutions usually present in a school setting.

By increasing their knowledge of mathematical investigations, the teachers were led to question their own teaching practices, a questioning that had positive results, as we can see by their worries and the changes made in their practices.

In conclusion, we can state that the factors that seem to act as triggers to improve teachers' professional knowledge are

- the open structure of the tasks and their investigation;
- the shared reflection in the course based on the investigative activities and teaching practices;
- the close relationship between the research and the teachers' education course; and
- the size of the group involved (small), which allows all the participants to hear the different points of view and ideas and to be heard, to act as their own agents of development.

Notes

1 This chapter is part of research conducted by Maiza Lamonato for her master's thesis, supervised by Cármen Lúcia Brancaglion Passos (see Lamonato, 2007).

2 A *bilboquet* is a toy made of two parts attached to one another by a string. When you do certain movements with your hands, it is intended that one end of the string, a sphere for example, will fit the other part.

References

Abrantes, P. (1999). Investigações em geometria na sala de aula. In E. Veloso, H. Fonseca, J. P. Ponte, & P. Abrantes, (Dirs.), *Ensino da geometria no virar do milênio* (pp. 51–62). Lisboa: Departamento de Educação da Faculdade de Ciências da Universidade de Lisboa.

Abrantes, P., Leal, J., & Ponte, J. P. (1996). Introdução. In P. Abrantes, J. Leal, & J. P. Ponte (Dirs.), *Investigar para aprender matemática* (pp. 1–4). Lisboa: Grupo Matemática Para Todos—Investigações na Sala de Aula (CIEFCUL) e Associação de Professores de Matemática.

Ernest, P. (1996). Investigações, resolução de problemas, e pedagogia. In P. Abrantes, L. C. Leal, & J. P. Ponte (Dirs.), *Investigar para aprender matemática* (pp. 25–48). Lisboa: Projeto Matemática para Todos—Investigações na Sala de Aula e APM.

Fiorentini, D. (2000). Pesquisando com professores: Reflexões sobre o processo de produção e ressignificação dos saberes da profissão docente. In J. F. Matos & E. Fernandes (Dirs.), *Investigação em educação matemática: Perspectivas e problemas* (pp. 187–195). Lisboa: APM.

Fiorentini, D. (2006). Grupo de Sábado: Uma história de reflexão, investigação, e escrita sobre a prática escolar em matemática. In D. Fiorentini & E. M. Cristóvão (Dirs.), *Histórias e investigação de/em aulas de matemática* (pp. 13–36). Campinas: Editora Alínea.

Fiorentini, D., Nacarato, A. M., & Pinto, R. A. (1999). Os saberes da experiência docente em matemática e a formação continuada de professores. *Quadrante, 8*, 33–60.

Freire, P. (1996). *Pedagogia da autonomia: Saberes necessários à prática educativa.* 31st ed. Coleção Leitura. São Paulo: Paz e Terra.

Freire, P., & Faundez, A. (1985). *Por uma pedagogia da pergunta.* 5th ed. Vol. 15 of *Coleção educação e comunicação.* Rio de Janeiro: Paz e Terra.

Lamonato, M. (2007). *Investigando geometria: Aprendizagens de professoras da educação infantil.* Dissertação (Mestrado em Educação), Universidade Federal de São Carlos.

Laville, C., & Dionne, J. (1999). *A construção do saber: Manual de metodologia da pesquisa em ciências humanas.* Porto Alegre: Editora UFMG/Artmed.

Mizukami, M. G. N. (2004). Aprendizagem da docência: Algumas contribuições de L. S. Shulman. *Revista Educação, 29*(2): 1-26. Retrieved 8 February 2011 from http:// coralx.ufsm.br/revce/.

Nacarato, A. M., & Paiva, M. A. V. (2006). A formação do professor que ensina matemática: Estudos e perspectivas a partir das investigações realizadas pelos pesquisadores do GT 7 da SBEM. In A. M. Nacarato & M. A. V. Paiva (Dirs.), *Formação do professor que ensina matemática: Perspectivas e pesquisas* (pp. 7–26). Belo Horizonte: Autêntica.

Ponte, J. P. (1994). O desenvolvimento profissional do professor de matemática. [Electronic version]. *Educação e Matemática, 31,* 9-20.

Saraiva, M. & Ponte, J. P. (2003). O trabalho colaborativo e o desenvolvimento profissional do professor de matemática. *Quadrante, 12*(2), 25–52.

Serrazina, L. (1998). *Teacher's professional development in a period of radical change in primary mathematics education in Portugal.* Dissertação de doutoramento, Universidade de Londres. Lisboa: APM.

Shulman, L. S. (1986). Those who understand: Knowledge growth in teaching. *Educational Researcher, 15*(2), 4–14.\

Chapter 14

The Role of Self-Regulated Learning in the Professional Development of Elementary Mathematics Teachers

Bracha Kramarski and Tali Revach

Introduction

Current reforms in the educational system have raised new goals for teachers' training concerning professional development (NCTM, 2000). In essence, these goals maintain that teacher training should not be limited to transmitting subject matter knowledge and pedagogical knowledge using predefined, fixed methods but find ways to construct knowledge through self-regulated learning (SRL), applying higher-order thinking skills. The ability to self-regulate learning is essential for teachers' professional development during their careers as well as for their ability to promote these processes among students. Research indicates that, if teachers are incapable of self-regulating their own learning, it will be impossible for them to develop these capabilities among their students (Putnam & Borko, 2000). Moreover, research has demonstrated that learners, across a wide range of ages, are not spontaneously self-regulated (Azevedo & Cromley, 2004; Kramarski, 2008; Kramarski & Michalsky, 2009), so it is necessary to implement explicit SRL support. Following this suggestion, our study addressed the following question. What is the effect of SRL support in professional programs on mathematics elementary teachers' development?

Supporting Self-Regulated Learning

SRL refers to a cyclical and recursive process that utilizes feedback mechanisms for learners to understand, control, and adjust their learning accordingly. The process involves a combination of four areas for regulation during learning: cognition, metacognition, motivation, and context condition (Schraw, Crippen, & Hartley, 2006). Cognition refers to strategies of simple problem solving and critical thinking. Metacognition refers to knowledge and control of cognitive skills. Motivation refers to learners' beliefs in their ability to learn, their intrinsic values for the task, and their interest level. Finally, context refers to evaluation and monitoring of changing task conditions.

A number of researchers have argued that several key factors support SRL through instruction: "embedding metacognitive instruction in the subject content matter to ensure connectivity; informing learners about the usefulness of metacognitive activities to make them exert the initial extra effort; prolonged training to guarantee the smooth and maintained application of metacognitive activity" (Veenman, Bernadette, Van Hout-Wolters, & Afflerbach, 2006, p. 9). These researchers have emphasized the generality of metacognitive skills and the importance of extensive practice, followed by explicit guidance in the classroom using the self-questioning strategy of WWWH (what, when, why, and how) that helps learners to select a specific self-regulatory strategy, approach, or response within learning (Kramarski & Mevarech, 2003; Schoenfeld, 1992). In particular, they have suggested the utility of structuring self-metacognitive questioning that focuses on learners' understanding of the task, awareness, and self-regulation of strategy application before, during, and after the learning task process.

Mevarech and Kramarski's (1997) IMPROVE method encourages learners to become involved in regulatory learning by using self-metacognitive questioning with regard to

1. comprehending the problem (e.g., "What is the problem/task?");
2. constructing connections between previous and new knowledge (e.g., "What are the similarities/differences between the problem/task at hand and the problems/tasks you have solved in the past?" and "Why?");
3. using appropriate strategies to solve the problem/task (e.g., "What are the strategies/tactics/principles appropriate for solving the

problem/task" and "Why?"; "When/how should I implement a particular strategy?"); and

4. reflecting on the processes and the solution (e.g., "Does the solution make sense?"; "How can I solve the task in another way?").

Generally speaking, research reported that supporting SRL with self-metacognitive questioning elicited positive effects on students' learning outcomes. However, little research exists in the field of in-service education to accurately determine the benefits and pitfalls of such a model in promoting in-service teachers' professional development in the mathematical domain.

Following this suggestion, the purpose of this study was threefold:

1. to investigate elementary teachers' professional development that was either exposed to IMPROVE metacognitive questioning support (SRL) or not exposed to SRL support (NS) with regard to pedagogical content knowledge of planning a lesson for an authentic problem-solving task;
2. to examine teachers' self-reported SRL skills with regard to motivation; and
3. to observe teachers' pedagogical and SRL processes of both groups in practice.

Method

Participants were sixty-four elementary teachers from twenty-seven elementary urban schools that comprise students of ages six to thirteen. Teachers in this level of education lack substantial mathematical and pedagogical content knowledge regarding new curriculum topics, such as algebra, and have limited experience teaching such topics. Teachers participated in a sixteen-hour course in a PD program to enhance their mathematical and pedagogical knowledge. The course was part of a three-year PD program sponsored by the Israeli Ministry of Education. One group of thirty teachers was assigned to the SRL support, and the other group of thirty-four teachers was assigned to no support. No significant differences between the two groups were found in the variables—years of experience teaching mathematics, mathematical and pedagogical content knowledge—that were assessed by the Ministry of Education measures.

PD Program: SRL versus NS

Based on Shulman's (1986) theoretical conceptualization of mathematical knowledge and pedagogical content knowledge, the purpose of the workshops was twofold: (1) to foster teachers' mathematical knowledge, and (2) to practise pedagogical content means for enhancing mathematical understanding. Each of the four workshops in both groups contained the same structure (NCTM, 2000). First, the instructor presented the lesson's subject and its contents to the teachers in the workshop. Second, the teachers practised the knowledge collaboratively in pairs. Practice was based on task solutions of various complexity that required comprehension of mathematical knowledge and pedagogical episodes and the analysis and evaluation of lesson plans or video-captured lessons. Each pair was asked to present their summary of the task solution and conclusions in front of the class using an overhead projector, addressing any difficulties that arose. And third, teachers were encouraged to participate in reflective discourse regarding interpretation of mathematical ideas and pedagogical events, understanding difficulties, and raising solutions and "appropriate" explanations for the problems. As part of their training, teachers were asked to conduct those lessons while practising various mathematical and pedagogical activities with their students and then to reflect on and discuss their experiences with their peers in the following meeting of the PD program.

The SRL teachers were exposed to SRL support based on the IMPROVE model as described before (Kramarski & Mevarech, 2003; Kramarski, Mevarech, & Arami, 2002; Mevarech & Kramarski, 1997). A detailed description of implementing tasks embedded with self-questioning in professional education can be found in Kramarski (2008) and Kramarski and Michalsky (2009).

Measurements and Procedure

Different kinds of measures were implemented during the professional development program. In this chapter, we focus on three kinds: teachers' pedagogical content knowledge, a self-reporting SRL questionnaire, and actual teachers' observations.

Teachers' Pedagogical Content Knowledge

Teachers' pedagogical content knowledge was assessed by their lesson plan on one of the authentic tasks that teachers solved in the workshop.

Planning requires teachers to create new components that demand high-order capabilities such as regulation, control, and evaluation of learning progress. The planning lessons were administered pre- and post-study. The lesson plan was analyzed according to three categories: (1) *task demands* (identifying the goal of the task, prior knowledge, and students' difficulties in solving the task); (2) *task design* (didactic ways of presenting the task, using different representations, learning environment); and (3) *teaching styles* (asking high-order questions, conclusions, and explanations). A sample lesson plan and scoring examples are provided in the appendix to this chapter.

For each category, teachers received a score of 3 (full use), 2 (partial use of two criteria), 1 (partial use of one criterion), or 0 (no use) and a total score ranging from 0 to 9. Two judges coded the lesson plans for all participants. Inter-judge reliability coefficients ranged from r = .86 to .88 for the three categories.

SRL Questionnaire

Motivation with regard to interest in the course and self-efficacy of teaching were assessed with a questionnaire of fourteen items based on the Motivated Strategies for Learning Questionnaire (MSLQ; Pintrich & De Groot, 1990) (e.g., "I believe that I have the ability to affect student achievement). Participants rated each item on a seven-point Likert scale, ranging from 1 (not at all true for me) to 7 (very true for me). Higher scores indicated a higher level of motivation. Cronbach alpha was 0.72.

Teaching in Practice

We observed the class practices of two teachers from each group: SRL versus NS. Teachers were selected randomly and observed during teaching the lessons that they had planned for the authentic task. Each lesson (ninety minutes) was videotaped, transcribed, and analyzed on pedagogical and SRL criteria: teaching for understanding and supporting SRL.

Procedure

Instruction began in the professional program at the beginning of the second academic semester and continued for four weeks (sixteen hours). The course comprised a unit on numbers and operations. Two pre- and post-measures (pedagogical content knowledge and self-reported SRL) were administered by the teachers' instructors in the classroom setting before the program started and at its completion. After completion of the post-tests, teachers were asked to teach their lesson plans in their upcoming lessons in the schools in which they were employed. Four teachers were randomly selected for videotaping. Participants were informed that these measures were part of a research study to determine the effectiveness of professional education programs.

Results

Teachers' Pedagogical Content Knowledge

Post-test ANCOVA results indicated differences between both methods of treatment on two criteria of lesson planning: *task demands* (M = 79.71; SD = 24.0; M = 63.42; SD = 30.60 respectively for the SRL and NS groups; F (1,61) = 4.78, $p < 0.01$; *effect size (ES)* = 0.60), and *teaching styles* (M = 91.30; SD = 20.64; M = 68.12; SD = 29.26 respectively; F (1,61) = 8.86, $p < 0.01$; ES = 0.93).

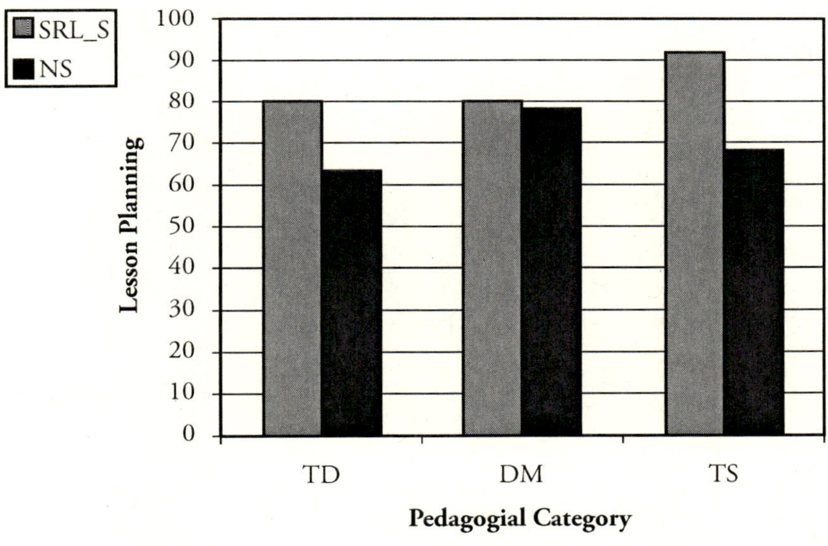

The SRL teachers were more aware of task demands (e.g., learning objectives), and they focused on activities for engaging students in learning such as asking for explanations and high-order questions. No statistically significant differences were found between the two groups on task design of didactic materials (M = 80.06; SD = 22.13; M = 78.41; SD = 25.03 respectively; F (1,61) = 1.86, $p > 0.05$; ES = 0.07).

Self-Regulated Learning

Two-way analyses of variance (treatments (2) by time (2) with repeated measures on the second factor) indicated significant interaction between treatments and time for motivation criteria ($F(1,61) = 4.79, p < 0.05$). We found that at the end of the study the SRL teachers exhibited significantly more interest in the course and self-efficacy in teaching such tasks than the NS teachers ($ES = 0.38$).

Teaching in Practice

As mentioned before, the transcripts were analyzed on two criteria: teaching for understanding, and supporting SRL.

Teaching for Understanding

We found that teachers focused on teaching for understanding (e.g., "We have to teach for understanding and not for technique") but differed in their pedagogical approaches to achieving this goal.

The SRL teachers focused on collaborative learning in small groups and reflective discourse in the whole class directed by the IMPROVE structure of the lesson: comprehending the problem ("What is the problem?" "Who can explain which savings plan is more profitable?"); constructing connections ("What is similar in the pattern of ... ?"); looking for appropriate strategies ("What is the strategy to find it?"); and reflecting on the solution ("Does the solution make sense?" "Can you suggest another solution?").

The NS teachers focused on a whole class discussion of problem solving. For example, one of the teachers read aloud the problem and explained the content (e.g., "The story of the task is familiar to you. The task includes the story of the problem and a graph that represents the data"). The other teacher did not guide the students to find the solution; rather, she asked questions but immediately gave answers ("What is the conclusion? You see, it depends on the interval of the time that we compare both plans") and summarized the conclusion ("Now you can write the correct conclusion in your copy books").

Supporting SRL

In general, the SRL teacher exhibited a more student-centred approach in class discussions (e.g., "You have to find the conclusions by yourself"; "Dan, what is your opinion?" "What is your conclusion?"), whereas the NS teacher exhibited a more teacher-centred approach to reaching conclusions (e.g., "Listen, we will find the conclusion together"). Moreover, we found that the SRL teacher more often directed students' learning explicitly with the IMPROVE self-questioning model, printed in the students' worksheets (e.g., "Did you ask yourself what is your strategy?" "How can we compare this task to the one from the previous lesson?"). The NS teachers used general comments (e.g., "Your friend obtained the answer; it doesn't mean that the solution is the correct one"), and when they used "why" questions they did not elaborate the answers but used a strategy of summarization.

Conclusion, Implications, and Future Research

Our findings indicate that supporting SRL with IMPROVE self-metacognitive questioning might be a vehicle for mathematics elementary teachers' professional development by promoting their pedagogical knowledge of teaching mathematics to strengthen their self-efficacy in implementing student-centred teaching approaches.

There are several possible reasons for the beneficial effects of SRL support on pedagogical content knowledge of mathematical problem solving and SRL. First, discourse on *why* and *how* questions seemed to foster teachers' understanding of task demands and pedagogical decisions. Second, the explicit opportunity to elaborate on different perspectives of problem solving, as both learners and teachers, appeared to lead teachers to focus more on deep understanding of task demands and on a student-centred teaching approach in their lesson planning. Third, the IMPROVE strategy engaged teachers in more student-centred learning, which seemed to strengthen the teachers' motivational beliefs about higher levels of interest and self-efficacy. And fourth, our study indicated that the NS teachers' simple sharing of methods, discussion of written work, and reflection on problems and solutions, which are all based on standards in mathematics learning, did not ensure that teachers understood how those standards benefit performance.

Our findings extend other findings regarding professional development models designed to deepen teachers' mathematical and pedagogical knowledge. For example, Farmer, Gerretson, and Lassak (2003) found that by explicitly reflecting on and discussing learning situations teachers noticed critical aspects that promoted their own learning and were motivated to change facets of their own teaching practices.

The study suggests integration in PD metacognitive programs of metacognitive knowledge and regulatory skills. Metacognitive programs such as the IMPROVE self-questioning guidance can help teachers to think systematically about the pedagogical steps that they need to take in their work, think about the deeper concepts and structures of disciplinary relations, and adjust appropriate teaching methods. When all types of questioning (comprehension, connection, strategic thinking, and reflection) are combined, probably constructed is a conceptual interconnected set of ideas that supports students' mathematical understanding and self-regulated learning approaches (Schraw et al., 2006).

We propose to examine additional effects of SRL support on various topics of elementary teachers' mathematical knowledge and pedagogical knowledge and to observe many teachers in class practice. In particular, we recommend that research investigate teachers' willingness to teach mathematics while using explicit SRL support such as the IMPROVE method. Moreover, future research is needed to connect these data directly to student understanding and achievement data.

References

Azevedo, R., & Cromley, J. G. (2004). Does training on self-regulated learning facilitate students' learning with hypermedia? *Journal of Educational Psychology, 96,* 523-535.

Farmer, J., Gerretson, H., & Lassak, M. (2003). What teachers take from professional development: Cases and implications. *Journal of Mathematics Teacher Education, 6*(4), 331-360.

Kramarski, B. (2008). Promoting teachers' algebraic reasoning and self-regulation with metacognitive guidance. *Metacognition and Learning, 3*(2), 83-99.

Kramarski, B., & Mevarech, Z. R. (2003). Enhancing mathematical reasoning in the classroom: The effect of cooperative learning and metacognitive training. *American Educational Research Journal, 40,* 281-310.

Kramarski, B., Mevarech, Z. R., & Arami, M. (2002). The effects of metacognitive training on solving mathematical authentic tasks. *Educational Studies in Mathematics, 49,* 225-250.

Kramarski, B., & Michalsky, T. (2009). Investigating pre-service teachers' professional growth in self-regulated learning environments. *Journal of Educational Psychology,* 101(1), 161-175.

Mevarech, Z. R., & Kramarski, B. (1997). IMPROVE: A multidimensional method for teaching mathematics in heterogeneous classrooms. *American Educational Research Journal, 34*(2), 365-394.

National Council of Teachers of Mathematics (NCTM) (2000). *Principles and standards for school mathematics.* Reston, VA: Author.

Pintrich, P. R., & De Groot, E. V. (1990). Motivational and self regulation learning components of classroom academic performance. *Journal of Educational Psychology, 82*(1), 33-40.

Putnam, R., & Borko, H. (2000). What do new views of knowledge and thinking have to say about research on teacher learning? *Educational Researcher, 29,* 4-15.

Schoenfeld, A. H. (1992). Learning to think mathematically: Problem solving, metacognition, and sense making in mathematics. In D. A. Grouws (Ed.), *Handbook of research on mathematics teaching and learning* (pp. 165-197). New York: Macmillan.

Schraw, G., Crippen, K. J., & Hartley, K. (2006). Promoting self-regulation in science education: Metacognition as part of a broader perspective on learning. *Research in Science Education, 36,* 111-139.

Shulman, L. S. (1986). Those who understand: Knowledge growth in teaching. *Educational Researcher, 15*(2), 4-14.

Veenman, M. V. J., Van Hout-Walters, B. H. A. M., & Afflerbach, P. (2006). Metacognition and learning: Conceptual and methodological considerations. *Metacognition and Learning, 1,* 3-14.

Appendix 14.1

The Lesson Plan

Nir and Yael decided to save each week the money they received from their parents. Nir received each day one NIS and Yael two NIS. Nir had saved four NIS from his previous plan, and Yael didn't have any savings.

1. Write the data in the following tables.
2. How much money will each child have after three, seven, and ten days? (The teacher prepared two tables that presented the amount of money they had saved along seven days.)
3. Draw an appropriate graph for each child. (The teacher prepared separate coordinate systems for each child.)
4. Are there equal data in both tables (days, amount of money)?
5. What is similar and what is different between the two graphs?
6. Which savings plan is more profitable? Explain.

Task Demands

The teacher referred to three categories in describing the task's demands:

1. The goal of the lesson: My lesson is designed for grade 5 students. Students should learn to organize data in a table and in a graph and to verbalize processes by using mathematical terms.
2. Prior knowledge: The lesson is based on students' knowledge about organizing data in a table and presenting data in a graph.
3. Learning difficulties: I expect some difficulties in administering the task in grade 5, such as organizing data in tables, presenting data in one graph, and finding generalizations.

This teacher received a score of 3 for this category of the plan. The planning referred to all three criteria of the task's demands.

Task Design

The planning referred to using representations (tables and graphs) but without didactic considerations for presenting the task, such as using different colours for the figure and learning environment as cooperative learning. This teacher received a score of 1 for this category of the plan.

Teaching Styles

The lesson asks for a high level of analysis (item 5), a conclusion, and an explanation (item 6). However, no specific emphasis was placed on high-order questions (e.g., "What is the similarity or the difference between the two lines in the graph?"). This teacher received a score of 3 for this category of the plan.

Chapter 15

Investigating and Writing in the Professional Development of Mathematics Teachers

Maria Teresa Menezes Freitas and Dario Fiorentini

Introduction

This chapter describes and analyzes the roles and contributions of both investigation and discursive writing in mathematics teacher education. After theoretically discussing this perspective in teachers' education, we describe the narrative inquiry (Clandinin & Connelly, 2000), which represents a methodological alternative to investigating the teachers' education process. We also present a narrative describing the personal and professional development of a student teacher who was observed over approximately three years, the period during which he became a teacher.

In this chapter, we take into account two studies, both developed at Unicamp (Fiorentini, 2004; Freitas, 2006). The last one concerns the domain of plane geometry and geometrical constructions, under the responsibility of the Department of Mathematics in the Mathematics, Statistics, and Scientific Computing Institute and focused on the mathematical content knowledge of the future teacher. The first study is related to the subject of mathematics teaching practice and teachers' supervised training (teaching practice) developed under the responsibility of the Department of Teaching Methodology of the Faculty of Education.

Investigation and Writing in Teacher Education

Didactical-pedagogical subjects have been studied due to their educative potential, using discursive writing and investigation, especially when these procedures are used and when we explore written discursive language registers of teacher learners (Fiorentini, 2004; Freitas, 2000).

Mathematics teacher education courses still maintain a teaching practice that values orality, explanation, repetition in long lists of exercises, and distribution of a systematic, formalized, and already built knowledge. Thus, the student doesn't reach this knowledge by doing explorations, inquiries, or readings. When writing appears in this context, it is usually technical and formal and it does not problematize the mathematical knowledge. This way of using and exploring writing in mathematics teachers' education does not contribute to professional development of teachers.

Faced with this problem, we were motivated to find out what happens during pre-service teachers' education when mathematics subjects also begin to privilege a process of reflective writing. Our proposal was to show that the intentional exploration of reading, researching, and writing in the process of teaching and learning mathematics is a potential way to expand future teachers' ability to understand and reflect on mathematical activity in the classroom, promoting in this way their personal and professional development.

According to Larrosa (1999), during the education process, the most important issue is not what the student learns but the intimate relationship that he establishes with the subject studied. In that sense, when the student deepens his studies, using themes in which he is interested and that respond to his expectations and possibilities, he is not reduced to a simple object of education. He becomes, at the same time, a product and an agent of his history and his own process of intellectual and human development (Fiorentini, 2004).

This conception of education, as well as the concept of experience in Fontana (2002) and Larrosa (1998), help us to improve the mathematics teaching processes that privilege the technical and formal reading-writing in mathematics to promote other ways of establishing relationships with mathematical knowledge: that is, this conception promotes knowledge and meaning provided by the teacher. Based on this conception, to obtain a genuine formative experience, mathematics teaching should consider an exploratory, communicative, and intersubjective practice, privileging the search for knowledge and the production of meaning about what is learned and taught.

Considering the possibility of research and discursive writing's contribution to a mathematics teachers' education course, we put ourselves in a position of curious investigators who tried to understand the educative potentialities that can be reached by the use of various communicative tools, especially writing.

The Practice of Investigating and Writing in the Courses

On the one hand, in the teaching practice course at Unicamp, the investigating and writing activities concerned, first, the pedagogical practice of the teacher educator (generally an experienced teacher who collaborates as a tutor or supervisor in the teachers' education program) and, second, the initial teaching practice of the student teacher. A day-to-day school study in this environment was carried out by pairs of students and developed by ethnographic observations in classes—registered in diaries—that were read, analyzed, and discussed by them and their colleagues (specialized in teaching practice).

On the other hand, the proposal of the plane geometry and geometrical constructions subject was to integrate geometry and language and to privilege activities in which students could write out their thoughts and ideas. Thus, the research of Freitas (2006) meets the interests of Professor Sandra, responsible for that curricular subject, for discursive writing activities, to produce meaning about the mathematical content knowledge explored in the course.

In this subject, students usually had to develop some activities with the support of a computer that would necessarily culminate in a written production, with pencil and paper, handed to the professor at the end of each class. The material demanded change according to the theme and needs of the activity. Different activities developed in the course included identifying the different steps in doing a geometrical construction; proving some statement; answering some question; writing a letter to an imaginary person (uncle or aunt) with no mathematics ability; and explaining how the student understood some construction, concept, or theorem.

Methodological Considerations: The Narrative Inquiry

To describe and interpret the real contributions of these approaches (developed in the geometry course and the teaching practice course) to the professional development of future teachers, we thought it necessary to establish not only a synchronic analysis but also a diachronic investigative movement. That is, we wanted to look at possible repercussions of such approaches in subsequent years, particularly when the student teachers would initiate their teaching practices in schools. We needed, therefore, a qualitative and interpretative approach that could respond to this demand, and we found it in the *narrative inquiry* proposed by Clandinin and Connelly (2000).

In the narrative inquiry, the participant shares with the researcher his personal and professional stories related to his actions or activities, providing in this way information and relevant traces of his professional development along the time period. As mentioned above, it is important to point out that in the research of Freitas (2006), in different moments and contexts, different components have allowed the development of many interactions: the interest of the professor responsible for the content explored in the course, the approach between the professor and the researcher—preserving an ethical position suitable to the research proposal—in an atmosphere of respect, friendship, complicity, and admiration.

The narrative inquiry represents, for Clandinin and Connelly (2000), a way of making sense of experience as well as a way of understanding it. So this kind of research assumes a tri-dimensional space constituted by the following dimensions: the "temporality," which involves past, present, and future; the "personal and social" interactions; and the setting where the narrated story happens.

This tri-dimensional metaphor of the narrative inquiry created by Clandinin and Connelly (2000) is similar to Bakhtin's (1988) notion of "chronotope." That term (literally, "time-space") was created by Bakhtin from Einstein's Theory of Relativity to express the inseparability of time and space, time being considered as the fourth dimension of space. In the narrative inquiry, the interdependence between "time" and "space" is essential to the comprehension of knowledge coming from experience.

Thus, on the basis of Clandinin and Connelly (2000) and Larrosa (1999), we tried to follow the developmental process of each participant and to be part of his or her movement, and to let us flow along his or her experiential development. On the other hand, in this research context, the narrative became the object of our study because the narrated stories were the focus of our analysis. On the other, the experiences developed in the course were also an object of study because it was through this narrative that we analyzed and explained the interpretations and meanings of the investigated experiences.

We followed the students of the geometry course during the first semester of 2003. Four members of this class were selected to revive and tell their stories at different moments, through autobiography (produced in the teaching practice course), interviews, and virtual and live conversations. To follow their personal and professional development, we organized the field notes and undertook repeated readings of these records to construct meanings of the experiences of each student teacher. Moreover, we made contact by email, conducted interviews at different moments (E_1, E_2, E_3), and had casual meetings during the two subsequent years (2004 and 2005).

For this chapter, we will just describe the PD trajectory of Frank Fitch, one of those four class members. We tried to capture and narrate his development by interweaving voices that helped us to understand the complexity of his subjectivity as well as his future as a researcher-teacher in the field of mathematics education.

Narrative of the Professional Development of Frank Fitch

Frank's wish to be a teacher stemmed from his childhood. The existence of teachers in his family (his mother and aunts) and the suggestion by some of his teachers to reflect on his choice made Frank choose the mathematics education course: "I have embraced mathematics, and I wanted to be admitted at Unicamp, because of the influence of some teachers" (Autobiography, Teaching Practice, 2003).

When Frank began to study geometry at Unicamp in 2003, he established links with different contexts along his path of development and identified discursive writing as something "new" and interesting: "The writing work was the major novelty for me; ... I could develop a real research project ... and not just deliver another work" (E_1).

At other moments, Frank evaluated writing as a kind of analysis and a catalyst for thinking. He also underlined the connection between practices of writing and analysis: "When we were writing, we have to analyze a lot, and sometimes we kept making demonstrations too but without using formal criteria. ... We had to think a lot. I have thought a lot" (E_1).

In 2003, Frank was also attending a teaching practice course. He developed a research project about mathematical investigations in the eighth grade and became interested in and aware of other possibilities for producing meanings for learning and teaching: "Reasoning can be visualized in other ways. ... I think it is possible to teach a content in a more meaningful way for the learner" (E_1).

The work that Frank has developed in the geometry course—combined with his other interactions, especially in the teaching practice course—stimulated his reflections regarding the relationship between theory and practice, and it was meaningful to him, in spite of the difficulties with his first contact with the Tabulae software: "Maybe it was one of the few opportunities I had, in which I could really link this course and this very institution with practice" (E_1).

Technological resources, combined with the mathematics content learning, seem to have given a distinguished status to this education for Frank. At the end of the semester, the proposal of a seminar in geometry was important in establishing the necessary links with his professional development. Talking with colleagues and with the teacher somehow made him feel like a teacher. So he started to rethink teaching and reconsider his conceptions: "I've changed even the way of thinking and facing teaching after that" (E_1).

One year after finishing the geometry and teaching practice courses, Frank added other learning opportunities, building a complex network of multiple relations in which mathematical investigations became prominent. Nowadays, his professional growth is also a result of his participation in a collaborative group of teachers of mathematics (the Saturday Group), a group that analyzes inquiries and writings about mathematics teaching practices at school. In this way, learning initially acquired in both courses served as a basis for continuing professional growth as a teacher. So Frank started to see himself as a teacher in a continuous process of development and as a researcher of his professional practice.

Although Frank found it difficult to be a teacher and a researcher at the same time, he recognized that discursive writing had been an indispensable factor in his professional development and an important resource for improving his teaching practice and the mathematical learning of his students.

In synthesis, the courses in *geometry* and *teaching practice* helped Frank, on the one hand, to realize the educative possibilities of non-symbolic and non-formal writing, on the other hand, to instigate theoretical and practical reflections on the process of learning and teaching mathematics mediated by writing.

The metacognitive potential that writing provides is brought out by Frank in a peculiar way: "Reflection is improved by writing. So, when you write, you reflect, and, when you read what you wrote, you 're-reflect'" (E_2). In this case, writing represented a strategy that provided a metacognitive reflection since it involved "thinking of thinking," and an ability to control his knowing process, when he investigated and wrote about mathematics teaching practice.

In a reflective movement of comings and goings, Frank revealed an important teaching knowledge and assumed that each reality demands a methodology that must be constructed by the teacher. He explained the necessity of improving his performance continually: "And, so, what the teacher has to do is to—he needs to give the best of himself, mobilizing multiple ways of explaining, ... and I learned this with the experience of writing that I had in the university pre-service course" (E_3).

In synthesis, Frank's professional development seems to have been supported and strengthened by a pre-service education developed from the perspective of professional development, mediated by investigating mathematics teaching practice through readings and writings about the process of teaching and learning mathematics through reflection and interaction with his colleagues.

Conclusion

The narrative of Frank Fitch's professional development reveals important aspects of the presence of discursive writing in his personal and professional growth. This presence appears in several contexts along his process of education: in his pre-service education, in his discursive and reflective writing practices in the mathematics learning and teaching process in geometry, and in his mathematics teaching practice after his graduation. In the teaching practice course, the discursive writing helped Frank to improve his perception and understanding of the complexity of school practices and his process of becoming a teacher in this context. However, the geometry course, associated with the use of information and communication technologies and the development of projects, provided metacognitive processes of mathematics learning and displayed other meanings of this subject (its concepts and processes), especially when the interlocutors were not familiar with formal mathematical language.

Besides, it became evident that educative experience, with discursive writing about mathematical content, in a teachers' pre-service course, can influence a teacher's future practice, as Frank also began to explore this peculiarity with his students. From this experience with his students, he developed a reflective and investigative stance toward his own practice, and the students' writing led him to understand their difficulties and points of view on mathematics teaching, pointing out directions in which practice could be modified or improved to achieve better learning results.

Frank's narrative also showed strong professional development during his pre-service course of mathematics teachers' education, characterized by the presence of reflective writing and investigation on mathematics pedagogical practice. During this trajectory, Frank (re)constructed some concepts and images about mathematics and its teaching. Furthermore, in his first years as a teacher, he also promoted an innovative and transformative way of teaching mathematics. His participation in a collaborative group of teachers (the Saturday Group) and his presentation of reports in congresses have revealed to him a teacher-researcher who contributes to the development of a teaching culture and the development of the professional field.

References

Bakhtin, M. (1988). *The dialogic imagination.* Austin: University of Texas Press.

Clandinin, D. J. & Connelly, F. M. (2000). *Narrative inquiry: Experience and story in qualitative research.* San Francisco: Jossey-Bass.

Fiorentini, D. (2004). A didática e a prática de ensino mediadas pela investigação sobre a prática. In J. Romanowski, P. L. O. Martins, and S. R. A. Junqueira (Dirs.), *Conhecimento local e conhecimento universal: Pesquisa, didática, e ação docente* (pp. 157-243). Curitiba: Champagnat.

Fontana, R. A. C. (2002). Exercícios de aprender: Os modos de ver, dos modos de ler, e de dizer por escrito o ofício de professor/a nas experiências de formação. In XI ENDIPE, Goiânia. *Anais* 1, CD-ROM, 12 pp.

Freitas, M. T. M. (2000). Estágio curricular em matemática na perspectiva de extensão universitária: Estudo de uma experiência na UFU. Dissertação (Mestrado em Educação), Faculdade de Educação, Universidade Federal de Uberlândia, Brazil.

Freitas, M. T. M. (2006). A escrita no processo de formação continua do professor de matemática. Tese (Doutorado em Educação), FE/Unicamp, Campinas, São Paulo, Brazil.

Larrosa, J. (1998). *La experiencia de la lectura: Estudios sobre literatura y formación.* Barcelona: Laertes.

Larrosa, J. (1999). *Pedagogia profana: Danças, piruetas, e mascarada.* Belo Horizonte: Autêntica.

Part 5

Reforms, Curricular Change, and Continuous Professional Development

Chapter 16

Factors Affecting the Implementation of Reforms in Secondary School Mathematics

Naomi Chissick

Background: Changes in Mathematics Teaching

During my work with mathematics teachers in the past seventeen years, I have come to appreciate the intricacy of their work on the one hand and its importance and impact on the other. The research described here stems from my experience as a teacher, head of a department, supervisor, and teacher educator in secondary mathematics education in Israel. My research looked at the factors that affect teachers in relation to reforms and innovations and the hierarchy of these factors in their view.

The Israeli school system is facing the challenges of reform in various ways. The Ministry of Education and Science and the Departments of Science and Mathematics Education in the universities and colleges have been working for many years to develop new teaching materials, educating new teachers and practising teachers in new teaching methods. However, these new approaches have penetrated few schools, and many teachers continue to teach the way they always have taught (Chissick, 2000). Thus, a study of the factors that affect the implementation of reforms seemed necessary.

At the same time, the study can be perceived fundamentally as an enquiry into adult learning paradigms. Knowles (1990) developed a theory of adult learning, or *andragogy,* that differs from other forms of learning in the following aspects:

- Adults are self-directed and expect to share the responsibility for decisions.

- Adults want to know why they need to learn.
- Adults want professional learning to be based on experience and be relevant to their immediate needs and practice.
- Adults want their professional learning to be collaborative and centred on peers.
- Adults want their learning to be problem centred and not necessarily content centred.

In this study, I looked at teachers' learning as a professional development process in a social setting. A project of implementation of reforms in mathematics teaching, focusing on didactic aspects (PIRM), was examined through the eyes of the teachers involved.

Project PIRM

Fullan (2001) differentiates between *restructuring,* which can be done by authorization, and *reculturing,* which deals with how teachers come to question and change their beliefs and, as a consequence, their practices. He claims that reculturing is needed nowadays but that it proves to be much more difficult than anticipated (p. 34).

In PIRM, a project that I developed for ORT (a network of colleges and schools for advanced technologies and sciences in Israel), with the aim of implementing reforms by creating a culture of teamwork, collaboration, and continuous professional development, we tried to promote reculturing with as little restructuring as possible.

The project was designed to answer the needs of mathematics teachers in the context of reforms and changes. It was conducted at a period when the mathematical curriculum was on the national agenda and when mathematics teaching in Israel suffered severe blows.

The project aims were

- to develop a culture of long-lasting professional learning and growth of mathematics teachers and teams;
- to encourage teachers and teams to try new concepts and methods in mathematics education;
- to enhance constant reflection on teaching and assessment practices; and

- to create communities of teachers (from different schools) that will take part in professional discussions and share materials and experiences for the benefit of all.

We tried to create a culture that will enable teachers to make lasting changes in their professional behaviour, including processes of continuous learning and growth. The core of the project was comprised of the guidance and support given to teachers to build a framework for collaboration and teamwork, to strengthen the status of the department head, and to reconsider their teaching methods. PIRM was based on

- constructivism as the leading knowledge paradigm;
- planning–action–reflection (Schön, 1983) in teaching and in teachers' learning;
- teamwork in the school and between schools; and
- in-school support for teachers and the department head by facilitators, once a week, for three years (Chissick, 2002, p.3).

It was assumed that professional development of teachers, both personal and as a team, should lead to improvements in students' mathematical thinking, learning schemas, and achievements. The suggested instructional approach for PIRM was derived from social constructivism (Jaworski, 1996; Vygotski, 1978) and the social practice theory (SPT) of Lave and Wenger (1991).

The strategy was under constant review to improve methods of implementation, find ways to cope with resistance, and promote regeneration of new approaches in mathematics education. The project was run for three years (1999-2002) in seven schools in the first year and thirteen schools in the third year.

Research Design and Methodology

"The biases of our careers, our personalities, and our situations constitute essential starting places for our research attention" (Wolcott, 1994, p. 408). The research and the research questions evolved from my deliberations while planning PIRM and from discussions with colleagues around the difficulties encountered during its operation.

The primary research question (**P**) was this: what are the factors affecting the implementation of contemporary mathematics teaching practices in secondary schools in Israel? The secondary research questions (**S1, S2**) were the following:

- What is the hierarchy of the factors affecting the implementation of reforms in secondary school mathematics in Israel?
- What is the place of the mathematics teacher's self-esteem in the hierarchy of factors affecting the implementation of educational innovations?

These three questions attend to gaps in knowledge that need to be studied to improve and rejuvenate projects such as PIRM.

The research combined socio-cultural perspectives and psychological perspectives on teacher change. This approach of taking account of and coordinating different perspectives was chosen because it seemed to be the most suitable to the problem at hand. The combined approach lies in the field of educational research and allows penetration into the area of processes of change in organizations.

The conceptual framework of the study emerged from the research questions, from my basic philosophical stance, and from extensive reading on the different research paradigms. It combined socio-cultural aspects, psychological aspects, and theories of change in educational settings.

The research (Chissick, 2005) was designed around the following theoretical and philosophical theories: hermeneutics as a leading philosophy and the constructivist learning paradigm as a leading construct, combined with naturalistic and inductive approaches and grounded theory. The interpretive and constructivist approaches used have roots in hermeneutics: both aim to understand human experience through the eyes of people connected to it.

The research followed cycles of data collection, interpretation, theory building, and more data collection—to verify the theories and uncover more information—from the perspective of the grounded theory approach (Glaser & Strauss, 1967). Data were collected from facilitators' reports and other project documents as well as interviews, observations, and questionnaires. A variety of methods was used for analysis, mostly qualitative. Case studies were done on two secondary schools, and vignettes from seven other schools were investigated.

Findings and Discussion

The factors affecting the implementation of reforms in secondary school mathematics education are numerous and varied. They include field conditions (time, administration's support, teamwork and peer support, external intervention, external assessment) and psychological conditions (teacher's self-esteem, belief system, and recognition of the need for change).

The study offers new insights into the roles that the different factors play in implementation processes and sheds light on their relative importance. This is a theme, to the best of my knowledge, that has not been studied before.

The hierarchy of factors was drawn from the coded data—observations, questionnaires, interviews, and written reports—combined with the grading of factors that teachers were asked to do at the end of PIRM, in a questionnaire distributed to all the teachers. The questionnaires were anonymous, but teachers had the option to fill in their names if they wished. Out of eighty-five teachers, sixty-four (75%) responded. The grading was between 1 (high) and 10 (low) and is presented in Table 16.1.

Factors 4 and 11 might be particular to ORT as a network. The data strongly suggest that the major factor supporting the implementation of reforms is the community of practice that surrounds the teacher, supports him or her, and shares the responsibility.

In a meeting held with one of the teams, near the end of the project, teachers said the following.

- I have learnt a lot from my colleagues, how to deal with the syllabus, what is important.
- I know what each of my colleagues is teaching.
- To this day, I use "tricks" that Mira and Ayala taught me.
- Here I learn. ...
- This is a warm group. I feel fantastic. I am not ashamed to ask. ...
- Everybody helped me. I arrived, a new immigrant, not knowing what to do.
- A strong team was formed. This is stable, there will be no retreat.
- We have discovered the strength and the possibilities in collaboration.

The head of the department and the external facilitator also play major roles as leaders of the team.

Table 16.1
Grading of factors affecting implementation of reforms

Factor	Rank (average grading)	Comments
Teamwork	1st (3.22)	In school A (an Arab school), the rating was significantly lower (6.125). The reasons seem to be cultural.
The HoD	2nd (4.64)	Schools A, M, and AE rated much lower (6.5, 6.43, 7.5). In School AE, the HoD was younger and less experienced than the other teachers.
The facilitator	3rd (4.72)	There is a significant difference between groups here. Schools NH, AE, and A rated 7.8, 8.5, and 6.125. All three were only two years in the project and have very mature teams. On the other hand, AH and NE had the same facilitator, very mature and knowledgeable, and their rating was high (1.33, 3).
Teacher development sessions (teamwork sessions)	4th (5.17)	Here School F, a vocational school, rated the sessions much higher than the others (2.33). This might be due to their unsatisfactory level of mathematical knowledge and agrees with their rating there.
The ORT supervisor and head of PIRM	5th (5.58)	School AE (new at ORT) gave a relatively very low rating.
Time	6th (5.82)	
Teachers' mathematical knowledge	7th (6.21)	See comment in 4th place.

(continued)

Table 16.1 (continued)

Factor	Rank (average grading)	Comments
The school's administration	8th (6.90)	
Shortage of suitable textbooks	9th (7.08)	
External examinations	10th (7.40)	Teachers realize that external examinations bring changes in content only, not in teaching practices.
Pupils' opposition	11th (8.13)	
Staffroom atmosphere	12th (8.67)	
Cooperation with teams from other schools	13th (9.33)	In PIRM, only the facilitators and HoDs had contacts with other schools.

Hargreaves (1994) elaborates on this phenomenon. He argues that some of the elements that make teamwork so important in reform situations are moral support, increased efficiency, reduced load, diminished uncertainty, increased capacity for reflection, opportunities to learn, and professional development. However, teacher educators should be aware that collaboration or controlled teamwork might cause complacency and encourage teachers to rely on others and relieve themselves from responsibility.

The psychological literature rates self-esteem as the most important element that rules our lives and guides our behaviour. However, mathematics teachers' self-esteem was found, in this study, to be no different from that of other teachers and its effect on implementation irrelevant. However, one can assume that a strong team increases its members' self-esteem.

Contextual factors were found to play a minor role in the process. A high workload, no set time or place for team sessions, and a lack of suitable teaching materials prevent the team from trying to develop new teaching practices. Evidence also shows that external, top-down changes lead to changes in curriculum but not in teaching practice.

Thus, the findings suggest that a strong, united, and supportive school team, with a department head who has leadership qualities, is an essential starting point for an effective process of change.

Conclusion

Socio-cultural factors appear to be the major force in the process of change, and contextual factors are important in enabling or hindering these processes. Psychological factors were found to be not as important as psychologists assume.

I would like to stress that the factors affecting implementation of reform cannot be separated into distinct categories. Therefore, I refer mainly to the factors themselves and not to their domains. However, I would like to present the findings as they inform two groups: the teachers and the education authorities.

Teachers tend to work in isolation. They are alone in their classes and usually have to take full responsibility for their pupils' achievements or failures. In addition, professional burdens and personal circumstances prevent many of them from joining professional courses. Thus, the process of continual learning and professional development is not an integral part of their agenda. Teamwork, as described here, supplies the opportunities for learning both content knowledge and didactics on the one hand and the umbrella of collaboration and shared responsibility on the other. Teachers should allow this to happen by devoting time to do it and opening their minds and hearts to learning and sharing.

Most of the findings also inform authorities and administrators (principals, teacher educators, and supervisors in particular). In line with the principles of andragogy mentioned earlier, teachers should be allowed to be active participants in the design and planning of their processes of professional development. In addition, field conditions should enable and enhance these processes.

However, further study is needed to better understand the *effectiveness and preservation of reform implementation* in education in general and in mathematics education in particular. Implementation of reforms takes a long time, and the results tend to fade as time goes on.

The results of this study strengthened my belief that PD strategies for mathematics teachers should follow the principles of social practice theory. The acquisition of knowledge, in the traditional sense of courses and seminars, should become part and parcel of the community-of-practice activities, in the teachers' natural setting, the school.

References

Chissick, N. (2000). In-service teacher training: Implementation of new teaching practices in high schools. Paper presented at the 9th International Congress on Mathematical Education (ICME-9), WGA7, Makuhari, Japan.

Chissick, N. (2002). Factors affecting the implementation of reforms in school mathematics. In A. D. Cockburn & E. Nardi (Eds.), *Proceedings of the PME 26th Conference* (pp. 249-256). Norwich: University of East Anglia Press.

Chissick, N. (2005). Factors affecting the implementation of reforms in secondary school mathematics. PhD diss., Anglia Polytechnic University, Chelmsford, UK.

Fullan, M. G. (2001). *The new meaning of educational change.* New York: Teachers College Press.

Glaser, B. G., & Strauss, A. L. (1967). *The discovery of grounded theory: Strategies for qualitative research.* New York: Aldine de Gruyter.

Hargreaves, A. (1994). *Changing teachers, changing times: Teachers' work and culture in the postmodern age (professional development and practice).* New York: Teachers College Press.

Jaworski, B. (1996). *Constructivism and teaching: The socio-cultural context.* Retrieved 5 March 2009 from http://www.grout.demon.co.uk/Barbara/chreods.htm.

Knowles, M. (1990). *The adult learner: A neglected species.* 4th ed. Houston: Gulf Publishing.

Lave, J., & Wenger, E. (1991). *Situated learning: Legitimate peripheral participation.* Cambridge, UK: Cambridge University Press.

Schön, D. A. (1983). *The reflective practitioner: How professionals think in action.* New York: Basic Books.

Vygotsky, L. S. (1978). *Mind and society: The development of higher psychological processes.* Cambridge, MA: Harvard University Press.

Wolcott, H. F. (1994). *Transforming qualitative data: Description, analysis, and interpretation.* Thousand Oaks, CA: Sage Publications.

Chapter 17

Teacher Development and Curricular Change: A Research Program

Dario Fiorentini, Rosana G. S. Miskulin, Regina C. Grando, Adair M. Nacarato, Cármen L. B. Passos, and Dione L. Carvalho

Investigating the Relationship between Teacher Development and Curricular Change

Professional teacher development and renewal of school curriculum have traditionally been treated as independent processes. Generally, curricular innovation and teacher education do not take, as starting and final points, the current school culture, the teaching conditions present in schools, and the complexity of pedagogical work in mathematics. In our opinion, this is the main reason why curricular reforms have not been able to achieve the expected transformations.

The studies developed by our research group GEPFPM[1] in past years in collaborative groups (gathering, in each group, professionals with different knowledge) showed that teachers' collective reflections on their practice had contributed to transforming the teachers into professionals (1) who are more critical of or reflective on their work, their knowledge, their students, and themselves; (2) who transform their practices with autonomy and conscientiousness; (3) and who produce knowledge in their professional field (Fiorentini, Freitas, Miskulin, Nacarato, & Grando, 2004; Fiorentini et al., 2005).

Teachers in Investigative Communities: Some Assumptions

The results of these researches led us to conceive teaching not only as a simple practice but also as a praxis in which theory and action are interwoven because teachers reflect and theorize all the time, negotiating possibilities between their classrooms and their lives in school and between their daily work with students and a larger movement of change and social equality. We projected, then, a way to promote the development of teaching practices and teachers' knowledge as well as an increase in their investigative communities. This seems to find an echo in Cochran-Smith and Lytle (1999, p. 292): "Constructing local knowledge is understood to be a process of building, interrogating, elaborating, and critiquing conceptual frameworks that link action and problem posing to the immediate context, as well as to larger social, cultural and political issues."

The collaboration among different professionals can help to construct teachers' identity and professional autonomy in order to become critical users of knowledge elaborated by others (Carr & Kemmis, 1987).

Our professional and research trajectory in Brazil has been devoted to articulating the problems and challenges of teacher education and the development of school curriculum. This has brought us to the conviction that university researchers, school teachers, and future teachers can learn together how to deal with school diversity toward a possible education of quality for the great contingent of less privileged students. So a possible education of quality, for each reality, is a collective construction and requires another conception of pedagogical work, of teacher development, and therefore of research.

In this conception, university teacher educators, school teachers, and future teachers constitute a community of learning and investigation in which the school teachers bring problems and challenges from their school practices and in which the teacher educators and future teachers act in response to the school teachers' demands. In this community, all the participants study, problematize, reflect, and investigate the complexity of mathematical school practices and negotiate possible changes in school curriculum (see Figure 17.1).

This problematization of the interrelation between teacher education and curricular change, in the current context, leads us, first, to assume anthropological conceptions of the introduction of "an extended vision of the possible content knowledge that must be included, explicitly, in the curricula,

Figure 17.1
Community of learning and inquiring

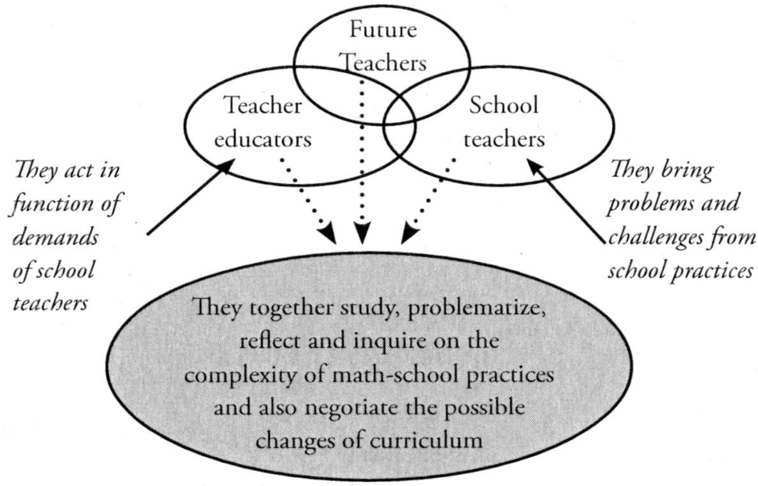

beyond the content knowledge which is considered academically relevant" (Gimeno, 1999, p. 174). Second, it leads us to assume a political and epistemological stance that consists of recognizing and taking account of the capacity of the teachers' community to promote professional knowledge, curricular change, and teacher development in collaborative and investigative ways. This second aspect focuses, as a starting point, on the problems and challenges of educational practices in schools. However, this is a complex social process with multiple expressions and dynamics which constructs itself over time and under certain conditions.

Our main research hypothesis is therefore to conceive of school and particularly teachers' work and school curriculum as cultural phenomena: that is, "as a crossing of cultures that provokes tensions, openings, restrictions and constraints in the construction of meanings" (Pérez Gómez, 2001, p. 12).

History of an Action and Research Program

In 2005, we started a coordinated action and research program involving four research groups in four Brazilian universities, the University Estadual de Campinas (UNICAMP) at São Paulo, the University Estadual Paulista

(UNESP) at Rio Claro, the University Federal de São Carlos (UFSCar), and the University of São Francisco (USF). This program's aim is to develop, coordinate, and systematize researches that analyze aspects related to the interrelation between curricular change and teachers' education or development. This analysis is conducted from a socio-cultural perspective that has, as background, didactical-pedagogical practices of an exploratory-investigative nature in a context of cultural diversity. The program therefore searches to understand this interrelation between curricular change and professional development of teachers and tries to develop a conceptual field giving a theoretical-methodological framework that could support the elaboration of projects articulating teachers' education and curricular development and researches in mathematical education.

The investigations developed in the scope of this program consider two perspectives:

1. *first-order researches,* composed of field or empirical researches carried out by teacher educators or master's and PhD students and by school teachers and graduate teachers who participate in collaborative groups;
2. *second-order researches,* composed of theoretical researches or systematization and/or qualitative meta-analysis of the studies and processes developed in the scope of the program (i.e., the material for analysis in this kind of research is constituted mainly by first-order researches and by reflection meetings, teachers' and researchers' analysis, and interpretation during realization and evaluation of their first-order research projects).

The development of this research program was confronted by the necessity to create a conceptual and theoretical framework during the practical and research process. It is why we saw the necessity and relevance of these second-order researches. We believe that this organic and dialectic movement between the first-order and second-order researches allows us to develop new directions, meanings, and issues for experiences and field researches. These experiences and researches will provide new elements and results for the second-order researches, from which other meanings and issues will be elaborated, and so on (Bakhtin, 2000). Such a movement can be understood looking at Figure 17.2, which illustrates the dynamics of the research process and the dialectic between the two perspectives of research.

Figure 17.2

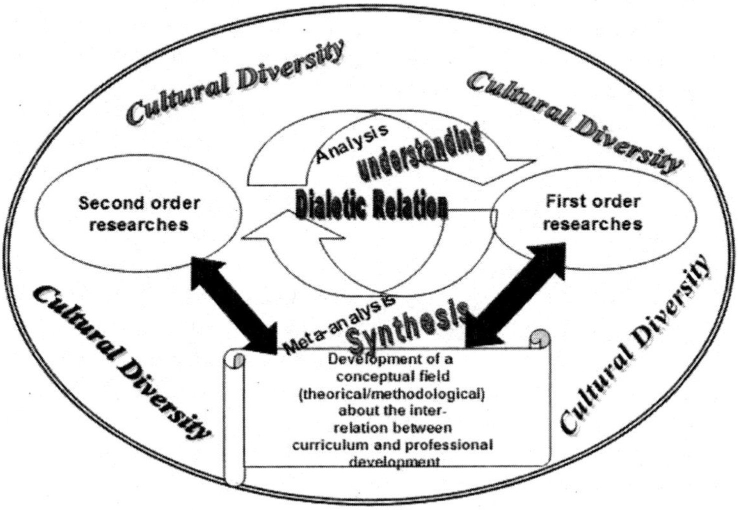

There is an interdependence between the second-order researches and the different contexts of the first-order researches which have been and will be carried out by participants, taking account of the specificities of each culture in which these researches are rooted: that is, the institutional culture, the teaching culture, and the school culture, among other ones. This movement does not imply an homogenization of the theoretical-methodological perspective to be used in understanding the investigated object. It is just the opposite. It provides another perspective, another understanding of the complexity of the processes involved in teacher education and curricular development. We have to underline that these orientations can be multiple but that they cannot be, due to the production conditions, any-one. In fact, among different participants, each culture mobilizes an *exceeding vision* (Bakhtin, 2000) that provides new meanings or interpretations.

Some Initial Results of the Research Program

Let us take, as material for analysis, five master's theses conducted in 2007 and developed by school teachers who participated in collaborative groups, sharing and discussing their researches. Although all these theses weave relations between teacher education and curricular practices, three of them (Andrade, 2007; Cristovão, 2007; Gomes, 2007) focus on exploratory-investigative practices developed in grade 5-8 classrooms; the two others (Farias, 2007; Lamonato, 2007) are concerned with practices and exploratory-investigative experiences developed in pre-service or in-service teacher education courses.

Let us consider, initially, the results obtained in the first set of research studies. The first research was done by Cristovão (2007) as an in-service mathematics teacher educator at a regional education department and as a teacher-researcher participating in a collaborative group at Unicamp. She realized the difficulty encountered by two teachers with cycle classrooms and recovery classrooms in mathematics. Then she formed a group with these teachers and other interested partners. This group was characterized as a collaborative group of first-order research-action. It became the object of analysis of her research, characterized by her as being second-order research-action (Elliot, 2000). The research's aim was to analyze the possibilities and contributions of exploratory-investigative activities for the inclusion of "weak" students in mathematics. In the first-order research-action, the teachers were able to discuss teaching alternatives and jointly plan and analyze tasks and activities of an exploratory-investigative nature, aiming at the mobilization and appropriation of concepts related to statistics and geometry. The analyses have shown that students considered "weak" in mathematics—by their teachers and by themselves—were able to have rich mathematical experiences. They could perceive themselves as being not only able to learn school mathematics but also able to produce their own ideas, arguments, and mathematical relations, thus increasing their self-confidence in mathematics. And the teachers themselves could experiment with and reflect on their practices, developing their autonomy in the production of school curriculum for heterogeneous classrooms and focusing on the practices of "weak" students as well as on the possibilities of inclusion of these students.

Gomes (2007), a teacher-researcher in a collaborative group at USF, developed an investigation with grade 5 and 6 classrooms in adult education (EJA), in which she aimed to analyze mobilization and production of mathematical knowledge generated in exploratory-investigative activities and the contributions of this type of activity to the teaching of mathematics in EJA classrooms. This analysis showed that such activities allow the teacher to give voice to the adults to express their ideas and thoughts mathematically, contributing to the development of an intellectual and critical autonomy and, consequently, to their social and educational inclusion. The researcher also showed the limits of these tasks due particularly to the "crowded classrooms, to the limited time to develop them in [the] classroom; to the culture of silence present in adult classes, to the physical and material conditions of the school; and to the students' preparation to work with this type of activity" (p. 172). In relation to her professional development, the teacher-researcher underlined that this research on her own practice, starting from opened tasks, helped to challenge herself in her daily practice, to mobilize knowledge, sociability, and critical and creative awareness, promoting continuous development of her professional identity.

Andrade (2007), also a teacher-researcher in a collaborative group at USF, investigated the pedagogical potential of virtual stories of mathematical concepts from the perspective of problem solving, from preschool to the end of elementary school. Interested in the written mobilization of mathematical concepts in these stories, and their written and oral productions, the study showed that these stories had motivated and mobilized the pupils, at all grade levels, to solve the problems presented in the stories. To solve them, they used pictorial representations at the preschool level and mobilized different strategies at the other grade levels. Moreover, the stories allowed the elaboration of new problems, the emergence of investigative tasks/activities, and the opening of space for pupils to attribute meanings to the contexts in which they were inserted. They also contributed to the emergence of many solving strategies that transcended the mobilization/production of mathematical concepts, approaching aspects related to their beliefs, values, and ideologies.

Regarding the two studies of the second group, let us consider first Lamonato (2007), a master's student who participated in a collaborative group at UFSCar. She analyzed a process of in-service training of four teachers who taught mathematics in kindergarden. This training process consisted of studies and discussions on the teaching of geometry in kindergarden. Its dynamics included experimentation, planning, and application of exploratory-investigative activities in geometry with further collective analysis. The author's intention was to investigate the learned facts and knowledge mobilized by teachers during the training process. The analysis confirmed that geometry is not a mathematical field present in their teaching practice. Regarding the exploratory-investigative activities, teachers developed an investigative and argumentative position during the acquisition of new knowledge about geometrical content. They recognized its importance when they realized that this content helped them to understand the place and value of geometry in kindergarden, as a way of interpreting and representing space. Teachers' actions developed in the classroom revealed their new ways of understanding the importance of children's drawings related to the games played and the changes in their ways of teaching.

The study of Farias (2007), a master's student who participated in a collaborative group at UNESP, Rio Claro, investigated the contributions of mathematical representations, as shown by undergraduate students in mathematics through the use of software, and the development of exploratory-investigative activities. These activities were developed in a differential and integral calculus course, part of the didactical-pedagogical education of future mathematics teachers on mathematics content knowledge. The results showed that, in this context, the flexibility of mathematical representations significantly helps the interpretation, visualization, and understanding of mathematical concepts. It also allows the future teacher to perceive the construction of knowledge that can be used in different ways: images, writings, logical models, and laws, which constitute the basis of these concepts. These shapes and the mathematical language played fundamental semiotic roles in the constitution and understanding of mathematical concepts.

Conclusion

These first-order researches were developed by school teachers who could count on the support and reflection of collaborative groups involving university mathematics educators, school teachers, and graduate teachers. The results showed that the teachers involved in these first-order researches presented evidence of professional development, such as

1. changes in the management and organization of the school curriculum, recognizing other effective possibilities to promote inclusion of pupils with learning difficulties in mathematics (this is the case of the exploratory-investigative tasks/activities, which break up a linear conception of the curriculum and allow changes in the beliefs and conceptions of pupils in relation to mathematics);
2. learning of new mathematical knowledge from the experience of this exploratory-investigative environment;
3. changes in attitudes in relation to mathematical knowledge and activity in the classroom, assuming a more critical and inquisitive

stance with pupils and opening up spaces for the production and negotiation of meanings; and

4. a more critical relationship with themselves and their practices at schools.

Briefly, teacher educators and teachers themselves recognize the importance of collaborative and investigative practices in facing the problems and challenges encountered in school practices. This collaborative movement of reflection on and investigation of the curricular practices, involving school teachers, teacher educators, and future teachers, metaphorically can be compared to a snowball effect: it begins small, but it can enlarge.

Note

1 GEPFPM is a research and study group on teachers' education located at the State University of Campinas (UNICAMP) that joins together researchers from four universities in Sao Paulo state, Brazil. This group develops field researches, meta-analyses, and state-of-the-art academic researches about education and professional development of mathematics teachers.

References

Andrade, D. O. (2007). *Contando histórias: Produção/mobilização de conceitos na perspectiva da resolução de problemas em matemática.* Mestrado em Educação, Universidade São Francisco, Itatiba.

Bakhtin, M. (2000). *Estética da criação verbal.* São Paulo: Martins Fontes.

Carr, W., & Kemmis, S. (1987). *Becoming critical: Education, knowledge, and action research.* London: Palmer Press.

Cochran-Smith, M., & Lytle, S. L. (1999). Relationships of knowledge of practice: Teacher learning in communities. *Review of Research in Education, 24,* 249-305.

Cristovão, E. M. (2007). *Investigações matemáticas na recuperação de ciclo II e o desafio da inclusão escolar.* Mestrado em Educação, FE/Unicamp, Campinas, São Paulo.

Elliot, J. (2000). *La investigación-acción en educación.* Madrid: Ediciones Morata.

Farias, M. M. R. (2007). *As representações matemáticas mediadas por softwares educativos em uma perspectiva semiótica: Uma contribuição para o conhecimento do futuro professor de matemática.* Mestrado em Educação, UNESP, Rio Claro.

Fiorentini, D., Freitas, M. T., Miskulin, R. G., Nacarato, A. M., & Grando, R. C. (2004). *Brazilian research on collaborative groups of mathematics teachers*. Paper presented at ICME-10, Copenhagen. Retrieved 5 March 2009 from http://www.icme-orgamosers.dk/tsg23/tesg23_abstracts/rTSG23025.

Fiorentini, D. , Miskulin, R. G., Megid, M. A., Brum, E. D., Gama, R. P., Melo, M. V., Reis, M. E. T., Grando, R. C., & Passos, C. L. (2005). Learning through collaboration from professionals with different knowledge. In *Proceedings of the 15th ICMI study on the professional education and development of teachers of mathematics,* Águas de Lindóia, Brazil. Retrieved from http://stwww.weizmann.ac.il/G-math/ICMI/strand2.html.

Gimeno, S. J. (1999). *Currículo: Uma reflexão sobre a prática.* Trad. Ernani Rosa. 3ª ed. Porto Alegre: ArtMed.

Gomes, A. A. M. (2007). *Aulas investigativas na educação de jovens e adultos (EJA): O movimento de mobilizar-se e apropriar-se de saber(es) matematico(s) e profissional(is).* Mestrado em Educação, Universidade São Francisco, Itatiba, São Paulo.

Lamonato, M. (2007). *Investigando geometria: Aprendizagens de professoras da educação infantil.* Mestrado em Educação, UFCar, São Carlos.

Pérez Gómez, A. I. A. (2001). *Cultura escolar na sociedade neoliberal.* Porto Alegre: ArtMed.

Chapter 18

Researching Effective Continuing Professional Development in Mathematics Education

Els De Geest, Marie Joubert, Rosamund Sutherland, Jenni Back, and Christine Hirst

Introduction

This chapter reports on the Researching Effective Continuing Professional Development (CPD) in Mathematics Education (RECME) project set up under the umbrella of the National Centre for Excellence in Teaching Mathematics in the United Kingdom (NCETM).

The NCETM was launched in 2006 with the broad aim of enhancing mathematics teaching and learning in schools, colleges, universities, and other organizations through high-quality continuing professional development. The NCETM is a major initiative funded by the Department for Children, Schools and Families (formerly the Department for Education and Skills) to enhance professional development for mathematics teachers in all education sectors in England. The NCETM was established in response to the recommendations of the Advisory Committee on Mathematics Education (ACME) and those made in the 2004 report of Professor Adrian Smith. The NCETM works with teachers of mathematics in two main ways: through a virtual presence on the Internet in the form of a portal (see http://www.ncetm.org.uk), and through a team of people working directly with schools, colleges, organizations, networks, the government, and everyone involved in mathematics education. All teachers and lecturers of mathematics, together with national and international organizations concerned with mathematics education, are considered stakeholders of the NCETM.

The philosophy of the NCETM is based on the notion that understanding and knowledge growth concerning the CPD of mathematics teachers are most valuable when they are co-constructed by teachers, researchers, and other stakeholders. This notion is described in detail by Triggs and John (2004), who use a model of multi-layered communities in which knowledge flow is omni-directional. They suggest that this approach can overcome the identified "gap" between research and practice (Weiss, 1998).

The NCETM is committed to research on professional development for teachers of mathematics and commissioned both the RECME project and a number of small-scale research projects. Its vision is that the findings of these projects will feed into the ongoing development of the NCETM to meet its aim of providing excellent advice, resources, and professional development for teachers.

The RECME project is ongoing and currently in the third of the four main phases. In this chapter, we describe in detail how the project was set up and developed and discuss the progress to date.

Laying the Foundations

Edwards, Sebba, & Rickinson (2007) observed that research design that builds on the co-constructing of knowledge in practice can be of several types. The RECME project bases its philosophical approach on that of the NCETM and extends toward a model of distributed leadership, which moves away from a "top down delivery" model of working:

> Taking this view [of distributed leadership], leadership is about learning together and constructing meaning and knowledge collectively and collaboratively. It involves opportunities to surface and mediate perceptions, values, beliefs, information and assumptions through continuing conversations. It means generating ideas together, seeking to reflect upon and make sense of work in the light of shared beliefs and new information; and creating actions that grow out of these new understandings. It implies that leadership is socially constructed and culturally sensitive. It does not imply a leader/follower divide, neither does it point towards the leadership potential of just one person. (Harris, 2003, p. 314)

The notion of distributed leadership implied calling on the knowledge and expertise of mathematics educators in research, policy, and practice. In the early stages, a Research Advisory Group (RAG) was formed, whose members (about twenty) were selected on the basis of their research reputations and on the educational sectors in which they conduct their work. The RAG designed the project in outline, first by discussing how they conceptualized the notion of professional development for teachers of mathematics, indicating the sorts of questions that they hoped the research would address, and then developing a set of aims. They put together a design for the project and appointed the project director, who joined their group.

A review of the literature concerning professional development for teachers of mathematics was commissioned, and the RAG played an active consultative and advisory role in the development of this review.

Research Questions and Aims

The first and possibly most important question that the RAG addressed was putting together a "definition" of CPD for teachers of mathematics. It was seen as important to be clear about what exactly they saw as CPD, to provide a starting point for the review of the literature. The following broad "definition" was developed: "CPD for mathematics teachers should stimulate teachers to re-think, to experiment, to make fresh distinctions and to probe those distinctions to explore how they are informative in enabling choices related to teaching and learning" (NCETM, 2009, p. 14).

The sorts of questions that the RAG developed were concerned with the context, content, and process (Harwell, 2003) of CPD. For example, they thought that it was important to understand the institutional and contextual settings in which programs of CPD took place, to take account of the content of the programs so that they could understand the kinds of professional capacity being developed, and they were interested in the forms of interactions afforded, the forms of experimentation evoked, and the stimuli provided. A further branch of questions was concerned with the short- and long-term changes that might occur as a result of CPD programs, with a focus on what counts as evidence of change and particularly what constitutes evidence of effective CPD.

This last question leads to the overarching aim of the RECME project, which is to investigate the interrelated factors that contribute to "effective"

CPD for teachers of mathematics. The other aims were based on the questions and were phrased thus:

- to characterize different types of continuing professional development for teachers of mathematics (to include both formal and informal experiences);
- to illuminate the types of evidence that could demonstrate that CPD is informing teachers' practices and students' learning;
- to investigate the influence of the NCETM portal on professional development for teachers of mathematics;
- to establish the roles of research in professional development for teachers of mathematics; and
- to inform future initiatives of NCETM.

Identifying the Landscape and Inviting Participation

It was considered important that there was wide consultation with all stakeholders in the early stages of the project, to draw out their views on CPD in general, in order to begin expanding joint understandings of what CPD is and to obtain a full, coherent, and detailed view of the CPD landscape in England.

To do so, in the first phase of the project (April 2007–July 2007), the Research Advisory Group, together with the NCETM regional co-coordinators, invited teachers, local authority advisers and providers of CPD for mathematics teachers to contribute to the database of CPD initiatives built via an online questionnaire on the NCETM portal. Potential contributors to the survey were alerted and invited to contribute through an advertisement in the national professional press (*Times Educational Supplement*), on the NCETM portal's home page, and through invitations emailed to members of professional and research associations (Association of Teachers of Mathematics, Mathematics Association, Association of Mathematics Education for Teaching, British Society for Research in Learning Mathematics).

The developing literature review also contributed to shared understandings of what is meant by CPD and which factors might contribute to "effective" CPD. This literature reviews research papers, policy documents, and inspection reports, but even at this early stage of putting the review together it became clear that the voice of the teacher was strangely quiet. So the literature can be seen as representing a "top down" perspective on the landscape of CPD, but the RECME project aims to include a stronger teacher voice.

Selecting the Sample

The RAG advised that the project should begin by studying a sample of about thirty initiatives. This, it was thought, would provide a wide enough range of types of CPD in the various education sectors. The relatively large number of initiatives would provide a broad overview of the nexus of factors contributing a joint understanding of CPD.

The sample was chosen from the 183 entries with information about past and present CPD initiatives logged on the portal. The selected CPD initiatives were contacted and invited to become part of the RECME project. The most important criterion for inclusion in participation was indication of willingness to engage in the research, working together with the research team to develop shared understandings. A further seven selection criteria were used to obtain a cross-section of the existing CPD landscape. These criteria represented variations in

- geographical areas based on English regions;
- structures in organization and settings, for example within school, across schools, led by government agencies, university involvement;
- subject knowledge of the teachers;
- educational sector, for example early years, primary, secondary, further education, adult education;
- criteria highlighted in the emerging literature review, for example type of CPD, such as lesson study, working on students' conceptual development; and
- ways of communication taking place within the initiative.

Above all, the initiative had to be ongoing during the academic year 2007-2008 to allow for "live" research to take place.

Not all criteria could be fulfilled from the CPD initiatives that were identifiable through the online survey. Further CPD initiatives that satisfied these "gaps" were sought by asking RAG members and other professionals in the field to suggest such initiatives.

Launching the Project

The project team invited CPD initiatives taking part in the research project to participate in a launch event in which they learned more about the project and took part in small group discussions. The purpose of these discussions was to enable the project team to draw on the participants' views to inform the development of a more detailed project design. The following questions were tabled in the first discussion:

- What is CPD?
- What is effective CPD?
- Where can we find evidence of effective CPD?

Thinking prompts were provided to evoke discussion. They were taken from statements from the online survey.

- Continually examining classroom practice in the light of new methods, research and implementing informed changes.
- Ongoing exploration and reflection on both the subject content of mathematics and the way in which pupils develop their understanding of mathematics in order to be increasingly effective in teaching it.
- The process of reflecting on and developing your professional skills.
- CPD is only effective if it takes the form of collaborative practice.
- I think this means that throughout our teaching careers we are still learners and there will always be ways in which we can increase our own subject knowledge so that it impacts on the students.
- Normally I would have expected CPD to involve attending courses, but I know you can go on studying by yourself and exchanging good practice with colleagues.
- Ways of ensuring that you do not get stale. Keeping interest alive. Ways to encourage people to experiment and exchange ideas.
- A way of linking teachers' Professional Development with the school development plan.

The second discussion session was based on the following questions.

- What kind of changes would you like to see in your students' learning as a result of this CPD?

- What would you consider convincing evidence of students' learning as a teacher?

Thinking prompts for these questions consisted of findings of what teachers considered signs of student learning in the IAMP project (Watson & De Geest, 2005).

- Students are more active in lessons, for example by participating in discussion, asking and answering questions, volunteering for tasks, offering their own methods.
- Students are more willing to share ideas with others: teachers, peers, the whole class.
- Students are showing more interest in mathematics, for example doing more homework, working on extended tasks, commenting positively in evaluation tasks.
- Students are more willing and able to tackle routine, non-routine, and unfamiliar tasks if they are offered to them—resilience and resourcefulness.
- Students act as mathematicians—looking for and expecting to find coherence in tasks; expecting mathematics to make sense.
- Students are doing better than expected or than past comparison groups.
- Students show improvements in behaviour and attendance.

The group discussions were all audio-recorded, and notes were taken by the research team (one member of the team for each group). The discussions will be collated, summarized, and published on the RECME portal, with an invitation to contribute to developing them as the full and coherent shared understanding of the community. This is in line with the philosophical underpinning of the project, which emphasizes the importance of inviting, and taking into account, the voices of the community.

Beginning the Research

The next phases concern refining the research questions and developing research instruments: how can we find answers to the research questions? Contributions from participants are sought by setting up an open discussion forum on the NCETM portal. The same questions used at the launch event are asked, and participants are alerted to the existence of this forum by email. Further contributions on the appropriateness and effectiveness of the research instruments and questions are collected during visits by the research team to five sample initiatives. Data and reports of visits are fed back to initiatives, and comments are requested.

In this way, the research instruments are built collectively, and through this process details of the nature of the data required and methods of data collection are developed by the project team in consultation with the RAG and the five initiatives studied (see above).

Collecting and Analyzing the Data

It is envisaged that data will be collected and analyzed in an iterative cycle. Members of the research team will visit each of the initiatives to collect data using semi-structured interviews and observations. Representatives of each initiative (chosen by the initiative) will be asked for comments on the data and will contribute to ongoing analysis of the data.

At the same time, and throughout the project, participants are also invited to report on ongoing professional development activities via the NCETM portal and to write articles, talk at meetings, write a blog, or wiki with the support of the research team.

This phase of the project will conclude with a twenty-four-hour residential summer school attended by the research team, representatives from the sample initiatives, some members of the RAG, and some regional members of the NCETM team. In this summer school, the data and analyses from the various initiatives will be presented for discussion, which will feed into the final analysis and findings from this phase of the project. Once again this way of working is in line with the philosophy backgrounding the project.

Final Phase

In the final phase of the project, selected initiatives will be studied in depth, producing rich case studies, presented as "thick" descriptions (Geertz, 1983). During this phase, teachers will be encouraged to engage with the project in the role of action researchers with the support of the research team. Gradually, the team will withdraw support while the teachers take on more responsibility for the research.

Concluding Remarks

The philosophy that values the co-construction of understandings greatly influenced the research design. We suggest that it helped to overcome the "gap" between research and practice by involving practitioners and researchers in design of the research, in analysis of the data, and in dissemination of the findings.

The project is characterized by a rigour that derives from ongoing scrutiny from the research community, participants in the research, and practitioners in mathematics education. This is because we share our progress on the NCETM portal, we welcome comments and criticisms, and we are eager to engage in dialogue concerning the research at any stage.

References

Edwards, A., Sebba, J., & Rickinson, M. (2007). Working with users: Some implications for educational research. *British Educational Journal, 33* (5), 647-661.

Geertz, C. (1983). Common sense as a cultural system. In C. Geertz (Ed.), *Local knowledge: Further essays in interpretative anthropology,* pp. 73-93. New York: Basic Books.

Harris, A. (2003). Teacher leadership as distributed leadership: Heresy, fantasy, or possibility? *School Leadership and Management, 23*(3): 313-324.

Harwell, S. H. (2003). Teacher professional development: It's not an event, it's a process. Center for Occupational Research and Development (CORD), Waco, TX. Retrieved 21 January 2004 from http://www.cord.org/uploaded-files/HarwellPaper.pdf.

NCETM. (2009). *Final report: Researching effective CPD in mathematics education (RECME).* Sheffield: NCETM.

Smith, A. (2004). *Making mathematics count: The report of Professor Adrian Smith's inquiry into post-14 mathematics education.* London: Department for Education and Skills, United Kingdom.

Triggs, P., & John, P. (2004). From transaction to transformation: Information and communications technology, professional development, and the formation of communities of practice. *Journal of Computer Assisted Learning, 20,* 426-439.

Watson, A., & De Geest, E. (2005). Principled teaching for deep progress: Improving mathematical learning beyond methods and materials. *Educational Studies in Mathematics, 58,* 209-234.

Weiss, C. (1998). Have we learned anything new about the use of evaluation? *American Journal of Evaluation, 19,* 21-33.

Conclusion

Education of Practising Teachers, Professional Life, and Development of Mathematics Teachers

Nadine Bednarz, Dario Fiorentini, and Rongjin Huang

The previous chapters give a rich portrait of alternative approaches to and experiences in mathematics teacher education aimed at supporting the professional development of teachers. These approaches and experiences try to be better aligned with teacher professional practices so as to be significant for them. The analyses in the different chapters help to clarify the characteristics of these various approaches, in particular how they articulate with practice, the roles played by the different participants (teachers, teacher educators, researchers), the underlying theoretical framework guiding the researchers, and the potentialities of these approaches to support teachers' learning *in, from,* and *for* practice.

To go further, we will first engage in a comparative analysis of these alternative experiences and approaches developed by researchers and teacher educators in different countries to shed light on some of their commonalities and differences. This reflection will question in particular the role of professional life in supporting professional development in practice, the nature of mathematics worked on, or the nature of collaboration between teacher educators/researchers and teachers. We will then present some outcomes suggested by these different contributions, highlighting the discussions held in the international topic study group (TSG28) in Monterrey, Mexico (ICME-11).

Different Experiences and Approaches Developed in Different Countries: Some Links

At the crossroads of alternative approaches and experiences developed in different countries in support of the professional development of mathematics teachers, we find a common foundation: a recognition that the professional development of teachers cannot be conceived of outside their practice and that it has to take seriously into consideration the teachers' points of view and practical knowledge to construct a significant PD experience. These experiences can take different forms, however, and be part of the professional lives of teachers. It is important to better understand, in that case, the informal learning process developed along the professional life of a teacher, specifically in relation to some crucial experiences encountered. Three different cases from China, the United States, and Portugal focus on some of these crucial experiences. Other approaches, articulated with practice, engage teachers in reflective activity about mathematics and its learning and teaching. We will present the similarities and differences that we saw among these different experiences and approaches.

Professional Development in Practice in China, the United States, and Portugal

We focus here on some experiences of teachers during their professional practices across different educational systems: teachers as subject leaders (Nunes), as mentors (Rhodes & Wilson), or as part of a system of promotion (Li, Huang, Bao, & Fan).

There are subtle differences among these experiences across different cultures even if they share similar terms. For example, even if all these experiences refer to teachers' leadership, it is described differently in the three educational systems considered: subject leader working in a school with other teachers (Nunes), expert teacher evolving through professional promotion (Li et al.), and mentor teacher working with novice teachers (Rhodes & Wilson). According to the description by Nunes, ome of the tasks of a subject leader include preparing and coordinating teachers' group meetings and project working sessions and supervising and evaluating colleagues' classroom practices. An important role for the subject leader is related to coordination and supervision. The subject leader is not necessarily a leading person in the content areas in terms of professional expertise and experience but a coordinator, keeping the subject group working together toward the support of all members. This process presents an important opportunity for the subject leader to develop her profession through collaboration with other teachers and this leadership experience.

Traditionally, mentor teachers, usually experienced teachers, mainly supervised student teachers or novice teachers and helped them to adapt to professional life. However, Rhodes and Wilson described a promising program in which mentors shifted their role from that of providing advice to that of working collaboratively with mentees. Thus, mentoring becomes a collaborative activity and a valuable learning opportunity for both mentors and mentees.

Li and his colleagues provided a holistic picture of a system of promotion in China. Expert teachers in that system are officially and socially recognized as subject leaders in terms of their expertise. They have the responsibility of supervising novice teachers (even experienced teachers) within or across schools, but they can also be the head of a teaching research group, director of a teaching department, or principal but not necessarily the formal subject leader. They are expected to play crucial roles in fostering other teachers' professional development by giving public lessons, delivering public lectures, and developing teaching materials. Thus, expert teachers in China are more oriented toward academic activity. In the Chinese experience, the leadership considered is wide and substantial and concerns the whole educational system.

This brief analysis of these three experiences points out some differences in how leadership is approached and indicates the need for further investigation of the nature of other professional experiences of teachers during their careers so that we can better understand their potential for professional development in practice.

Approaches to and Experiences of Professional Development in Brazil

The six Brazilian studies presented in this book emphasize approaches and experiences aimed at supporting the professional development of teachers. These approaches and experiences are based on reflection, collaboration, and inquiry and focus on practices of teaching and learning mathematics in different contexts. They are founded on the assumption that teachers can construct knowledge in and from practice. Their knowledge and their socio-cultural practices are the starting and arriving points of continuous professional development. All these experiences recognize the "capacity of the teachers' community to promote professional knowledge, curricular changes, and teachers' development in a collaborative and investigative way" (see, e.g., Fiorentini et al.).

From the point of view of in-service education, Brazilian educative experiences involve the participation of school teachers and teacher educators/researchers (see, e.g., da Costa; Ferreira & Miorim; Passos & Lamonato) and sometimes of pre-service teachers and postgraduate students (see, e.g., Nacarato & Grando) in a collaborative work environment. In this context, school teachers share and discuss experiences and difficulties in their teaching, with the support of teacher educators, read and discuss papers on new tendencies in teaching, and elaborate methodological alternatives (see, e.g., Ferreira & Miorim), often characterized by an exploratory-investigative approach (see, e.g., Nacarato & Grando; Passos & Lamonato). School teachers implement these alternatives in their classrooms, collect information, register students' classroom interactions on video, and create oral and written narratives of participants/researchers. Then these materials are taken by all members of the group as objects of analysis. Emerging from this process, new knowledge in mathematics education is developed, and pedagogical situations, for teaching and learning mathematics in schools, are collectively achieved.

Although most of these group studies are developed within the university, other groups can be constituted within the school. This is the case of the group reported by da Costa that involved a partnership between university and schools and the participation of different professionals, non-specialists in mathematics teaching (teachers, principal, pedagogical coordinator), who played different roles (learners, teachers, teacher educators, and researchers) in the process of teaching and learning mathematics and statistics, using computers, in elementary schools.

From the point of view of pre-service education, under the aim of professional development, Brazilian educative experiences consist of introducing the future teacher to the process of reading, reflecting, inquiring, and writing his own practice of learning mathematics in a specific discipline (e.g., geometry), and in other teacher-training disciplines, as well as his own practice of becoming a mathematics teacher within the school. The focus of the study relative to the teacher-training disciplines is, initially, the pedagogical practice of another teacher and evolves toward a study of his own initial teaching practice (Fsee, e.g., reitas & Fiorentini). Another experience of pre-service education consists of introducing the future teacher to a collaborative group, in which he participates as a novice member in a community of teachers and teacher educators who study, problematize, reflect, and investigate the complexity of mathematical school practices while also negotiating possible changes to the school curriculum (see, e.g., Fiorentini et al.; Nacarato & Grando). According to these authors, this approach is a potential way to extend the future teacher's ability to reflect and understand a mathematical activity in the classroom while promoting his personal and professional development as well as contributing to the development of his professional community.

This characterization of the Brazilian experiences, emerging from the analysis of the six studies considered in this book, can be linked to approaches developed in other countries, for example in Canada around collaborative research (Bednarz, 2004; Bednarz, Poirier, Desgagné, & Couture, 2001) or in Canada and England around investigative activities (see Jaworski; Proulx; Suurtamm & Vézina). We will return to these commonalities and differences below.

Approaches to Professional Development in Different Countries

To characterize the nature of programs developed for educating practising teachers (continuous professional development), Barbara Jaworski distinguished the nature of teachers' participation in them:

- courses or summer institutes led by teacher educators, where teachers are considered students;
- research and (or) development programs led by teacher educators, where teachers are participants;
- collaborative research in learning and teaching, where teachers are partners in research with teacher educators/academics; and
- research programs in which teachers explore their own practices and in which they are researchers.

This reference framework related to the nature of teachers' and teacher educators' participation helps us to see some similarities but also differences among the approaches developed in this book. In some cases (Beatty & Bruce in Canada; Kramarski & Revach in Israel), teachers are at some moment of the process considered as students; this is the case in the research/PD model developed for algebra (Beatty & Bruce) during the teaching of lessons by researchers or in Kramarski and Revach during the PD program centred on self-regulated learning. Teachers are also considered as participants; this is the case in the research/PD model (Beatty & Bruce) during assessment of the lessons elaborated by the researchers. It is a completely different way to support professional development that we find in a collaborative process in which teachers and researchers are engaged in the co-construction of knowledge related to practice (Bednarz 2004; Bednarz et al., 2001; Bednarz, Maheux, & Barry, 2007). In other cases, action research is involved, and teachers act as researchers (see, e.g., the model developed by Fiorentini et al. in Brazil for the elaboration of curriculum by teachers). We also found, in some cases, a more complex process of participation of teachers, their role changing during the approach: from participation of teachers as students, to teachers who teach in the classroom, to teachers who teach other teachers (teacher educators), to researchers (see, e.g., da Costa in Brazil). This process also involves in some cases different status (see, e.g., Nunes from teachers to teacher leaders; Li et al. from apprenticeship-based practice in school to teaching-based research group; Rhodes & Wilson from teachers to mentor teachers who support pre-service or novice teachers).

The role of teacher educators/researchers in the approaches can also help to point out some important differences. In some cases, the teacher educator/researcher is in a position of expert in the research domain investigated (see, e.g., Beatty & Bruce for algebra and patterning; da Costa at the beginning of the process, in the introduction of software). In other cases, the teacher educator is in the position of facilitator (see, e.g., Rhodes & Wilson), trying to support the collaborative work of practising teachers and pre-service teachers. The researcher is sometimes engaged in the co-construction with teachers of new knowledge related to practice (Bednarz, 2004; Fiorentini et al.). In this collaborative context, researchers are interested in the investigation of teachers' development trajectories or a better understanding of some components of teachers' practices, taking into account the practitioner's voice.

A certain sensibility toward the practical contexts of practitioners seems necessary in these cases (on the notions of practical sensibility and theoretical sensibility, see Desgagné & Bednarz, 2005). A certain conception of practitioners as socially competent actors who have practical knowledge thus underlies several of these approaches. This sensibility toward the practical contexts of practitioners has led some researchers to question the nature of the mathematics worked on in the usual models of teacher education.

Questioning Mathematics Worked On

As several researchers have pointed out in recent studies, how mathematics teachers engage with a mathematical situation is quite different from how mathematicians would engage with it (Ball & Bass, 2003; Bednarz & Proulx, 2009; Moreira & David, 2008). The gap between the mathematical experiences of (pre-service and in-service) teachers in their teacher education programs and those of actual in-school practice has been documented (Proulx & Bednarz, 2008), raising fundamental questions about the current orientations of mathematics teacher education programs. These studies suggest a rethinking of the mathematical experiences and learning opportunities that teachers are exposed to through their professional formation (at the pre- and in-service levels). Three chapters of this book focus on this question and present alternatives related to this mathematical preparation.

In the chapter by Proulx, an approach to in-service education is developed that attempts to offer teachers experiences better aligned with their mathematics teaching practices. As developed elsewhere (Proulx & Bednarz, 2008), this initiative focused on (1) the development and enrichment of teachers' knowledge of the mathematics that they teach (i.e., school mathematics) and (2) their immersion in an engaging practice as mathematical "doers."

In the chapter by Sánchez, mathematical events are reflected through notes of reflection, which constitute the core of interactions between teachers from different schools and different levels. The note of reflection is in fact organized around a fictitious school situation designed around a mathematical question or problem, inspired by real teaching experiences. Some links can be established with mathematical events in the classroom which become the occasion to engage teachers in a mathematical reflection.

In the chapter by Murphy, focusing on pre-university mathematics, the mathematical investigation takes the form of a course conceived in collaboration among Mathematics in Education and Industry, universities, and schools, built in a flexible manner to give teachers the opportunity to choose areas of study. If we are close here to the usual approach of mathematical content knowledge, a certain articulation with practice is present in the conceptualization of the course.

The mathematical preparation of teachers (at the pre-service and in-service levels) is an important field of investigation of the worldwide community of researchers, opening up new conceptualizations of "using mathematics in teaching" and possible significant alternatives in pre-service and in-service education (Hache et al., in press).

In the different chapters of this book and during the topic study group and the collective discussions held at ICME-11, important ideas and key issues were apparent. We try, in the second part of this conclusion, to synthesize them.

Main Ideas and Issues

Challenges Related to Collaborative Work

Collaborative work is an important characteristic of several experiences/approaches presented in this book (see, e.g., the Brazilian experiences and approaches in Nunes; Rhodes & Wilson; Sanchez). What is the nature of this collaboration?

Collaboration between Teachers or between Teachers and Teacher Educators

As teacher educators/researchers, we have to be careful about the assumption (implicit or explicit) that "collaboration transforms teachers." It is not automatic. There are in fact different types of collaboration (see, e.g., the work developed from the socio-constructivist and Vygotskian perspectives in Garnier, Bednarz, & Ulanovskaya, 1991) and reflection involved in this process of collaboration. Some answers to the question of collaboration (What do we mean by collaboration? In what sense is it different from cooperation?) were proposed during the discussions in Monterrey. In cooperation, students or teachers engage in solving a problem or completing a task collectively. In collaboration, there are the ideas of

- working together (collaboration is related to professional work and not to a precise task);
- sharing the same goals;
- contributing to the practice investigated with different perspectives, experiences, and professional knowledge; and
- opening up room for possibilities.

Some similarities among the different approaches presented can be pointed.

- They focus on the work of teachers.
- They are organized around specific mathematical content (geometry, patterning and algebra, statistics) or processes (proofs, inquiries).
- A process of collaboration is engaged between persons with various experiences.

But there are also important differences.

Differences among Collaborative Approaches

These differences concern *the status accorded to the collaboration* in the approaches presented.

- A collaboration *designed a priori:* The model proposed by the researchers/teacher educators involves collaboration among partners at some moment of the process (see, e.g., Beatty & Bruce on the assessment of lessons in patterning and algebra).

- A collaboration *conceived as a process in which the different partners are engaged:* Different approaches engage partners in a collaborative process around a common object of investigation related to practice (see, e.g., da Costa on the teaching of statistics; Nacarato & Grando on learning experiences in geometry; Nunes on collaborative groups of teachers sharing and reflecting on their own practices; Rhodes & Wilson on practising teachers and pre-service teachers using teachers' practices to stimulate reflection).
- A collaboration *emerging during the process:* The researcher follows and analyzes its development (see, e.g., Ferreira & Miorim on analysis of the evolution of the collaborative work of mathematics teachers).

These differences also concern *the actors involved in this collaboration.* This component is important to consider because, at the core of this collaboration, depending of who is involved, a mediation among different cultures of practice is involved in the co-construction among the different partners (Bednarz, 2004; Desgagné, Bednarz, Couture, Poirier, & Lebuis, 2001). In the different chapters, we found

- a collaborative process involving teachers with various experiences (Ferreira & Miorim; Nunes; Rhodes & Wilson; Sanchez);
- a collaborative process involving practitioners from different cultures of practice (principal, pedagogical coordinator, teachers, subject leaders) (see da Costa); and
- a collaborative process involving teachers and teacher educators/researchers (including pre-service and in-service teachers, graduate students) (see De Geest, Joubert, Sutherland, Back, & Hirst; Nacarato & Grando; Suurtaam & Vézina).

Related to the roles of the different partners in the process of collaboration, the question of power is also a key issue to consider.

Implicit Power Relations in Collaboration

Are there, in a non "top-down" design like those presented in the different chapters, implicit power relations that can present challenges to collaboration? It is important to examine such implicit power relations. Different ways to approach this question were explored in the group discussions.

We can relate implicit power relations to the concept of "relation to knowledge." The different partners in this collaboration occupied what we can call a certain "position of knowledge." The central question, asked by Jaworski in her chapter, is "How is knowledge distributed between teachers and teacher educators/researchers?" In the different chapters, this position and relation to knowledge are not always the same.

In some cases, researchers, based on their previous research, occupied the position of expert (Beatty & Bruce; Kramarski & Revach). Lessons on patterning and algebra are built by researchers, and teachers have to appropriate these lessons, their points of view being related to assessments of these lessons in their classes.

In other cases, we can speak of

- knowledge shared by teachers and researchers (Ferreira & Miorim);
- co-generative action (da Costa);
- co-construction of new knowledge related to practice, researchers and practitioners contributing to this elaboration of new knowledge in activating resources from their experiences and practices (Bednarz et al., 2001); or
- production of knowledge by teachers (in the tradition of critical action research, we enable the empowerment of practitioners; Fiorentini et al.)

The question of implicit power relations was also related to that of "taking an inquiry stance." When collaboration between teachers and researchers focuses on inquiry as a tool for developing classroom teaching and learning, as did the work of Jaworski, we are effectively confronted by this question of power. Researchers are engaged naturally in an inquiry stance, but teachers at the beginning of the process are not necessarily in the same stance (what we mean by inquiry is not necessarily clear for them). It is interesting to see that teachers themselves are beginning to speak of inquiry during the process, an indication that something is changing among power relations.

Investigative Activities and Reflection

Starting from chapters centred on the exploration of investigative activities by teachers in mathematics (Fiorentini et al.; Freitas & Fiorentini; Passos & Lamonato; Suurtamm & Vézina), the discussion in the topic study group focused on the role of reflection in the learning process and the professional development of teachers.

The Importance of Reflection

Some commonalities among the different approaches/experiences presented in this book can be pointed out:

- an importance accorded to reflection;
- an approach in which teacher educators do not tell teachers what to do but allow them to develop in their own ways, asking their own questions;
- a supportive environment;
- a process of collaborative work; and
- "modelling" ways of working with teachers that they can use in a similar way with their students.

But there are also differences. In some cases, teachers are involved in mathematical investigative task(s), and the chapters seem to refer to different types of reflection.

- What do we mean by reflection?
- What are the different kinds of reflection we refer to?
- What are the relationships between this process of reflection and teachers' professional development?

Related to the issue of teacher change, the role of exploratory-investigative tasks seems to be one important trigger. Reflection is basically a cognitive process but can be unconscious. However, it is fundamental in the process of change. There is a dialectical relationship between reflection and change.

Relationships between Reflection and Teacher Development

The different chapters suggest that not just any kind of reflection can be educative or support the professional learning of teachers. Some chapters refer to different kinds of reflection: written reflection (Freitas & Fiorentini); oral and written narratives (Nacarato & Grando); self-regulated learning (Kramarski & Revach); and working and reflecting within the context of practice (Li et al.; Nunes; Rhodes & Wilson).

For instance, for practising teachers, written reflection on teaching and learning mathematics seems to help them perceive relations in their practices that they did not perceive previously and appears as a meta-reflection (Freitas & Fiorentini). The narrative seems to provoke changes in the way people understand themselves and others (Nacarato & Grando).

For future teachers, written reflection on mathematical teaching practices in school contributes to a better understanding of the complexity of these practices. It also contributes to a better understanding of their own process of becoming a teacher (Freitas & Fiorentini). On the other hand, this written reflection, having as a reference the process of becoming a teacher, promotes metacognitive reflection on mathematics and other mathematical meanings, particularly in relation to future teachers' conceptions and beliefs about school mathematics, teaching, and learning (Freitas & Fiorentini).

On the Learning of Teachers and Teacher Educators

Another dimension considered during the discussion is the learning of teacher educators or researchers, who are themselves learners. What do they learn while working *with* teachers? What do they learn *from* teachers? These questions are raised by Jaworski and taken as an important issue to consider.

Considering the learning of not only teachers but also teacher educators/researchers represents *a shift of perspective*. Teachers and teacher educators are learners in practice; they actively learn from each other to improve their own practice (Proulx; Rhodes & Wilson). New knowledge is built by researchers in this process: systematizing their experiences (Nacarato & Grando), developing new theoretical models and deep thinking/insight (De Geest et al.).

Working with teachers also requires *a shift of perspective in the way we consider the evaluation of learning developed by teachers.*

- How can we "evaluate" (together) the growth of teachers?
- What really happens after the professional development experience? Does professional development still go on? How do teachers reappropriate the meanings of these experiences? (In Fiorentini et al., e.g., do teachers continue to develop the curriculum and change their practice out of the PD context?)
- What knowledge is developed in practice? What are the new insights developed in these experiences (Li et al.; Nunes; Rhodes & Wilson)?

Conceptualization of Professional Development

Some theoretical questions underlay the functioning of this topic study group. What do we mean by teachers' professional development in mathematics? Which conceptualizations emerge from research in mathematics education about professional development? What are their different theoretical and epistemological bases?

An innovative way to approach this conceptualization was initiated by De Geest and colleagues and is consistent with the foundations of collaborative approaches developed by researchers and teacher educators. This project is centred on the co-construction among teachers, teacher educators, and researchers of the definition of effective continuous professional development, engaging a wide community in mathematics education in a viable characterization of continuous professional development, which directly concerns actors involved in it.

Some important characteristics are identified by this way of approaching the question of professional development, a complex and multi-faceted project, a way to take account of the complexity of teaching and teachers' education.

- It focuses on examining various methods of professional development, opening rather than closing the conceptualization.
- It sheds light on the important point that professional development is continuous.
- It tries to bridge the gap between research and practice.
- It gives voice to teachers.
- It is connected to teachers' learning in practice and teachers' evidence of students' learning (the inside voice, not the outside voice).

This conclusion is an attempt to compare the different approaches developed in this book to support professional development. This comparative analysis allows us to better understand the affiliations but also the differences among these new avenues explored in mathematics teacher education (at the pre-service and in-service levels). Moreover, through reflection on some important issues discussed in the topic study group, we raise some questions and aspects to consider in the analysis of these approaches and experiences in the future.

Note

We want to thank all the members of the topic study group (TSG28) held at ICME-11 in Monterrey, Mexico, who contributed to making a very productive and rich group. The ideas developed during the discussion have been reconstructed to form the basis of the second part of this conclusion.

References

Ball, D. L., & Bass, H. (2003). Toward a practice-based theory of mathematical knowledge for teaching. In E. Simmt & B. Davis (Eds.), *Proceedings of the 2002 annual meeting of the Canadian Mathematics Education Study Group* (pp. 3-14). Edmonton: CMESG.

Bednarz, N. (2004). Collaborative research and professional development of teachers in mathematics. In M. Niss & E. Emberg (Eds.), *Proceedings of the International Conference on Mathematics Education,* 4-11 July, Copenhagen. Copenhagen: Cédérom.

Bednarz, N., Maheux, J. F., & Barry, S. (2007). Context-based professional development in mathematics education: A collaborative research project. In T. Lamberg (Ed.), *Proceedings of the 29th Annual Conference of the North American Chapter of the International Group for the Psychology of Mathematics Education,* 25-28 October, Lake Tahoe (pp. 788-795). Lake Tahoe: Cédérom.

Bednarz, N., Poirier, L., Desgagné, S., & Couture, C. (2001). Conception de séquences d'enseignement en mathématiques: Une nécessaire prise en compte des praticiens. In A. Mercier, G. Lemoyne, & A. Rouchier (Eds.), *Sur le génie didactique: Usages et mésusages des théories de l'enseignement* (pp. 43-69). Collection Perspectives en Éducation et Formation. Bruxelles: Éditions de Boeck.

Bednarz, N., & Proulx, J. (2009). Knowing and using mathematics in teaching: Conceptual and epistemological clarifications. *For the Learning of Mathematics, 29*(3), 11-17.

Desgagné, S., & Bednarz, N. (2005). Médiation entre recherche et pratique en éducation: Faire de la recherche "avec" plutôt que "sur" les praticiens. *Revue des sciences de l'éducation, 31*(2), 245-258.

Desgagné, S., Bednarz, N., Couture, C., Poirier, L., & Lebuis, P. (2001). L'approche collaborative de recherche en éducation: Un rapport nouveau à établir entre recherche et formation. *Revue des sciences de l'éducation, 27*(1), 33-64.

Garnier, C., Bednarz, N., & Ulanovskaya, I. (Eds.) (1991). *Après Vygotski et Piaget: Perspectives sociale et constructiviste: Écoles russe et occidentale.* Bruxelles: Éditions de Boeck. (2nd édition 2003; 3rd édition 2008).

Hache, C., Proulx, J., & Sagayar, M. M. (in press). Synthèse du GT#1—formation mathématique des enseignants: Contenus et pratiques. In A. Kuzniak & M. Sokhna (Eds.), *Actes du colloque Espace Mathématique Francophone EMF-2009*

/ *Enseignement des mathématiques et développement: Enjeux de société et de formation*. Dakar, Sénégal: Université Cheikh Anta Diop. CD-ROM.

Moreira, P. C., & David, M. M. (2008). Academic mathematics and mathematical knowledge needed in school teaching practice: Some conflicting elements. *Journal for Mathematics Teacher Education*, *11*(1), 23-40.

Proulx, J., & Bednarz, N. (2008). The mathematical preparation of secondary mathematics schoolteachers: Critiques, difficulties, and future directions. Paper presented at ICME-11 in TSG-29, July, Monterrey, Mexico. http://tsg.icme11.org/tsg/show/29.

Contributors

Jenni Back is an associate professor (senior lecturer) in mathematics education based at the University of Plymouth. After many years of teaching mathematics in primary and secondary schools, she studied for her PhD at King's College London and is an experienced teacher educator working at both initial teacher education and masters level. Her research interests are focused on professional development for teachers of mathematics; comparing approaches to teaching mathematics in Hungary and England; classroom interaction and talk; and communication about mathematics through digital media.

Jiansheng Bao is a professor of mathematics education at East China Normal University. His research interests include teacher education, psychology of mathematics education, and international comparisons of mathematics education.

Dr. Ruth Beatty is an Assistant Professor teaching Mathematics Methods in the Faculty of Education at Lakehead University (Orillia Campus). Her research focuses on how children learn complex mathematical concepts, particularly in the domain of "early algebra" that has resulted in a comprehensive curriculum for students in grades 4 to 10. Other research interests include models of professional development that link research and practice, and the integration of technology in mathematics teaching.

Nadine Bednarz Nadine Bednarz is a professor emeritus at the Université du Québec à Montréal (Canada). Her research interests include mainly collaborative research in mathematics education and professional development of teachers, mathematical preparation of teachers at both pre- and in-service levels, teachers' mathematics mobilized in their practice, and transitional issues in mathematics teaching and learning. She is a member of the GREFEM (Groupe de recherche sur la formation à l'enseignement des mathématiques). She has published several books and papers in mathematics education in the international scientific community.

Catherine D. Bruce is an associate professor of mathematics education at Trent University (Canada), where she teaches mathematics methods courses. Her research program focuses on pre-service and in-service teacher efficacy, models of teacher professional development, and student achievement.

Dione Lucchesi de Carvalho has a PhD in mathematics education from Unicamp, Brazil. Currently, she is a professor in the Faculty of Education at Unicamp, São Paulo. She teaches in undergraduate programs in mathematics and in postgraduate programs in education. She is a member of the research group PRAPEM (Pedagogical Practices in Mathematics). Her research interests are on pedagogical practice in mathematics, adult education, and statistics education.

Naomi Chissick has been the head of mathematical studies for ORT (a network of schools and colleges for advanced technologies and sciences) in Israel for the past eighteen years. She has worked mainly with secondary schools and teams of teachers in in-service education. Her main goal has been to enhance teaching, with an emphasis on mathematical thinking, problem solving, and learning through inquiry. She has been involved in the implementation of reforms in mathematics education and initiated didactic changes in teaching practices.

Nielce Meneguelo Lobo da Costa is a professor in the Mathematics Education Program at the Bandeirante University of São Paulo, Brazil. She works with both pre-service and in-service mathematics teachers at the secondary level. She has completed her PhD in education (curriculum) at the Pontifical Catholic University of São Paulo, becoming part of the research group FORCHILD. Her main research interest is Information and Communcation Technology (ICT) in mathematics teachers' education and professional development.

Yadong Fan is a mathematics teacher at Suzhou High School (China). He is a master teacher with the highest rank for school teachers, equivalent to the rank of full professor in an university. His research interests include effective teaching of mathematics and instructional design. He is the author of a series of mathematics textbooks for high school.

Ana Cristina Ferreira is a professor in the Department of Mathematics at the Federal University of Ouro Preto (UFOP), Brazil. She has been the coordinator of the postgraduate course in mathematics education at the UFOP for over four years. She obtained her master's and PhD degrees in mathematics education at the State University of Campinas, Brazil. Her main research interests are mathematics teachers' education and professional development, collaborative work, history of mathematics and mathematics education, research methodology in mathematics education, and pedagogical practice of mathematics.

Dario Fiorentini is a professor in the Faculty of Education at the State University of Campinas (Unicamp), Brazil. He is currently a coordinator of the postgraduate courses in education at Unicamp. He leads the research groups PRAPEM (Pedagogical Practices in Mathematics) and GEPFPM (Study Group in Mathematics Teachers' Education). He has a master's degree in mathematics and a PhD in education from Unicamp. His research interests are mainly in teacher education and professional development in communities of inquiry. He has published papers and books on teacher education and professional development, collaborative research and action-research, and state-of-the-art and qualitative meta-analysis of research in mathematics education.

Maria Teresa Menezes Freitas is currently an associate professor in the Faculty of Mathematics at the Universidade Federal de Uberlândia (UFU), Brazil. She is also the coordinator of the Mathematics Education Area in the Faculty of Mathematics and director of the Centre for Distance Education of the university. She has completed her PhD in mathematics education at Unicamp, São Paulo, becoming part of the research group GEPFPM (and has many publications on teachers' education). She has been the coordinator of the Mathematics Laboratory at Uberlândia Federal University for over ten years. Her main research interests are mathematics education and teachers' professional development, narrative as a method of research, and Information and Communcation Technology (ICT) in teachers' education.

Els De Geest taught for many years in both comprehensive and independent secondary schools in the United Kingdom and worked as a mathematics consultant for local government and tutor in teacher education. She was the director of the Researching Effective Continuous Professional Development (CPD) in Mathematics Education (RECME) project for the NCETM (UK). She has been involved in several research projects at the University of Oxford. She currently works at the Open University (UK). Her research centres on the process of making sense and its application in the teaching and learning of mathematics.

Regina Célia Grando has a PhD in education from the State University of Campinas, Brazil. She is a professor in the postgraduate program in education Stricto Sensu, working in the Mathematics, Culture, and Pedagogical Practices research program of São Francisco University, Brazil. She is currently a coordinator of the postgraduate courses in education at São Francisco University. Her research interests are in the field of mathematics teachers' education, problem solving, and the use of games in teaching mathematics. She coordinates with Adair Mendes Nacarato a collaborative group focused on the teaching of geometry (Grucogeo).

Christine Hirst has been involved in educational research for the past six years. The main focus of her research has been in the area of graduate mathematical education, and she has worked with schools (primary and secondary) through the study of a number of other educational programs. She has also worked with organizations and businesses studying the numerical and mathematical skills in terms of employability. Her other research interests include the transfer and management of knowledge.

Rongjin Huang is an associate professor of mathematics education at Middle Tennessee State Unviersity, USA. He worked as a high school mathematics teacher and a mathematics educator in China for many years. He is a member of Learner's Perspective Study, a well-known international classroom research project with more than sixteen participating countries. His research interests include mathematics classroom research, mathematics teacher education, and comparative mathematics education. He completed several research projects and published more than thirty journal papers, edited books, and book chapters both nationally and internationally. His recently published book is *Mathematical Tasks in Classrooms around the World* (with Shimizu, Kaur, & Clarke, Sense Publishers, 2010). He organized and chaired several topic study sessions at various regional and international professional conferences such as EARCOME 3 and 4, PME 30, 31, and 32, and ICME-10 and -11.

Barbara Jaworski is a professor of mathematics education at Loughborough University in the United Kingdom and a visiting professor at the University of Agder in Norway. She is the director of research in a Mathematics Education Centre that focuses on research on mathematics teaching and learning in the university. Her research is focused on mathematics teaching and its development, with a special focus on communities of inquiry and developmental research. She was editor of the *Journal of Mathematics*

Teacher Education for six years and is one of the editors of the *Handbook of Mathematics Teacher Education* (2008).

Marie Joubert worked as a teacher of mathematics and information technology for over twenty years before studying for a PhD at the University of Bristol (United Kingdom). While working on her PhD, and after its completion, she was involved in various research projects focusing on school mathematics and the use of computer technology in the classroom and in research. She is currently based at the University of Bristol as a researcher.

Bracha Kramarski is an associate professor in the School of Education and is currently the head of Mathematical Teacher Education at Bar-Ilan University in Israel. Her teaching focuses on mathematics education, metacognition, computer-based learning, and teachers' professional development.

Maiza Lamonato is a mathematics teacher in elementary school and a teacher educator at public schools in the city of Ribeirão Preto, Brazil. She has a master's degree in education and is a regular PhD student in education at the Federal University of São Carlos, Brazil. Her present research is on the professional development of teachers teaching mathematics and on exploratory-investigative activities in mathematics. Her research interests are also on teacher education and teaching practice.

Yeping Li is a professor of mathematics education at Texas A&M University, USA. He is to be the Head of Department of Teaching, Learning, and Culture starting on September 1, 2011. Is is also the holder of Claude H. Everett, Jr. Endowed Chair in Ecucation at Texaqs A&M University. His research focuses on issues related to mathematics curriculum and teacher education in various education systems and how factors related to mathematics curriculum and teachers shape effective classroom instruction that is valued in different cultures. His research projects have been supported by various funding agencies, such as the Spencer Foundation. His work has been published extensively in many journals internationally. He has been serving as an associate editor of the *Journal of School Science and Mathematics* and a guest editor of other journals, including the *International Journal of Educational Research*, *ZDM* and *The International Journal on Mathematics Education.* He organized and chaired many group sessions at various prominent national and international scientific conferences, such as ICME-10 in 2004 and ICME-11 in 2008.

Maria Ângela Miorim obtained a master's degree in mathematics and a PhD in mathematics education at the State University of Campinas, Brazil. As professor at Campinas-São Paulo, she is working on mathematics teacher education (pre-service) and postgraduate studies in mathematics education. She coordinates (in collaboration with Antonio Miguel) the research group HIFEM (History, Philosophy, and Mathematics Education), affiliated with CEMPEM (Circle of Study, Memory, and Research in Mathematics Education). Her research publications are in the field of relations between history and mathematics education.

Rosana Giaretta Sguerra Miskulin has a PhD in mathematics education from Unicamp, Brazil, where she also did postdoctoral studies at the College of Education, as a FAPESP researcher, in technology and education. She was a visiting research professor at the University of New Mexico, United States. Currently, she is a professor in the Department of Mathematics at the State University of São Paulo, Rio Claro, UNESP—Brazil. She teaches in undergraduate and postgraduate programs, where her role is a coordinator of the Postgraduate Program in Mathematics Education. Her research interests are in teaching practice, professional development, mathematical learning/teaching process, and information and communication technology (ICT), with an emphasis on the use of software in the classroom and distance learning.

Bernard Murphy was a teacher of mathematics to eleven to eighteen year olds in schools in England and Tanzania for twelve years before joining Mathematics in Education and Industry (MEI; see www.mei.org.uk) to design and run the Teaching Advanced Mathematics course. He is now the MEI programme leader for continuing professional development.

Adair Mendes Nacarato has a PhD in education from the State University of Campinas, Brazil. She is a professor in the postgraduate program in education Stricto Sensu, working in the Mathematics, Culture, and Pedagogical Practices research program of São Francisco University, Brazil. Her research interests are in the field of mathematics teacher education and mathematical practice. She leads the research group HIFOPEM (Histories of Formation of Mathematics Teachers) and coordinates with Regina Célia Grando a collaborative group on the teaching of mathematics (Grucomat).

Cláudia Canha Nunes has completed her master's degree in the didactics of mathematics at the University of Lisbon (2004), with a thesis on assessment as regulation of the teaching and learning process in mathematics. She has been a mathematics teacher since 1997 in a middle school of Lisbon. She has also been the coordinator of the Department of Mathematics and participated in this school in several projects in mathematics. She was part of the European project on Professional Development of Teachers as Researchers. She participates regularly in national professional meetings and international conferences. She is developing her doctoral dissertation in the didactics of mathematics and is a member of the coordinating committee of the investigation working group of the Association of Teachers of Mathematics in Portugal.

Cármen Lúcia Brancaglion Passos has a PhD in mathematics education from Unicamp, Brazil. In 2008, she took part in a postdoctoral research program in the Faculty of Sciences at the University of Lisboa. Currently, she is a professor in the Department of Teaching Methodology of the Federal University of São Carlos, Brazil. She teaches in undergraduate and postgraduate programs, where her role is that of coordinator. Her research interests are in teaching practice, teacher education, and teachers' professional development in teaching mathematics.

Jérôme Proulx is a professor of didactics of mathematics in the Department of Mathematics at the Université du Québec à Montréal(Canada). He works in the mathematics teacher education program. His research interests concern the mathematical preparation of teachers at both pre- and in-service levels. He is a member and the coordinator of GREFEM (Groupe de recherche sur la formation à l'enseignement des mathématiques).

Tali Revach graduated from the School of Education at Bar-Ilan University in Israel. Her thesis was on mathematics teachers' professional development.

Ginger Rhodes is an assistant professor in the Department of Mathematics and Statistics at the University of North Carolina, Wilmington, where she teaches mathematics education courses for pre-service and in-service teachers. Her research interests include teachers' understanding and use of students' mathematical thinking in instructional practices.

Mario Sánchez teaches at the National Polytechnic Institute, Mexico. He has been a mathematics teacher educator at the mathematics education program PROME, CICATA-IPN, in Mexico. His research interests include issues of interaction, communication, and reflection in online teacher education. He has completed a PhD in mathematics education research at Roskilde University in Denmark.

Rosamund Sutherland is a professor of education at the Graduate School of Education at the University of Bristol (United kingdom). She has taught mathematics in secondary schools and further education colleges. Her research centres on teaching and learning with Information and Communcation Technology (ICT), with a particular focus on mathematics education.

Christine Suurtamm is an associate professor of mathematics education at the Faculty of Education of the University of Ottawa (Canada). Her research interests include teachers' facilitation of inquiry-oriented mathematics learning and new forms of assessment. She is the director of the Pi Lab: Promoting Inquiry in Mathematics Education, a research facility for analyzing video data.

Nancy Vézina is a consultant at the Centre for University Teaching at the University of Ottawa (Canada) and a researcher in the Pi Lab. Her research interests include a variety of topics connected to elementary mathematics teaching, with a particular interest in the connections between student understanding and teacher practice. Along with Christine Suurtamm, she forms a bilingual research team with projects in both English-language and French-language contexts.

Patricia Wilson is a professor in the Mathematics Education Program at the University of Georgia, USA, where she works with both pre-service and in-service mathematics teachers at the secondary level. Her research interests centre on mathematics teacher development and the preparation of mathematics teacher educators. She is currently the project director of the Center for Proficiency in Teaching Mathematics, a collaboration with the University of Michigan.

Index

MARQUIS
Marquis Book Printing Inc.
Québec, Canada
2011